Tactical Communications for the
Digitized Battlefield

For a listing of recent titles in the *Artech House Information Warfare Series,* turn to the back of this book.

Tactical Communications for the Digitized Battlefield

Michael J. Ryan
Michael R. Frater

Artech House
Boston • London
www.artechhouse.com

Library of Congress Cataloging-in-Publication Data
Ryan, M. J. (Michael J.)
    Tactical communications for the digitized battlefield/Michael Ryan, Michael R. Frater.
        p.   cm. — (The Artech House information warfare library)
    Includes bibliographical references and index.
    ISBN 1-58053-323-x (alk. paper)
    1. Communications, Military.   I. Frater, Michael R.   II. Title.   III. Series.
UA940 .R93   2002
358'.24—dc21   2002019679

British Library Cataloguing in Publication Data
Ryan, Michael
    Tactical communications for the digitized battlefield.
    (Artech House information warfare library)
    1. Communications, Military   2. Electronics in military engineering
    I. Title   II. Frater, Michael R.
    623 . 7'3

    ISBN 1-58053-323-x

Cover design by Yekaterina Ratner

International Standard Book Number: 1-58053-323-x
Library of Congress Catalog Card Number: 2002019679

10 9 8 7 6 5 4 3 2 1

Contents

Preface

Early commanders were in intimate contact with their troops. Battlefields were small—tiny in comparison with those of today—and commanders could direct their troops by using their voice or their physical presence to influence the outcome of the battle. As commanders embarked on more adventurous campaigns, the size of their forces prevented close-quarter command, and commanders had to remove themselves somewhat from the battlefield so that they could obtain a larger view than was possible while standing among the troops. Effective command, therefore, required commanders to occupy a convenient hilltop from which to oversee own-force and adversary dispositions. While this new position increased situational awareness, it also increased the distance between the commander and subordinates, which increased the time to convey orders and created a delay in their transmission that had to be accommodated for by commanders in their plans.

The increased distance also required commanders to develop communications systems to assist in passing orders. Early systems were courier-based, making use of messengers to carry orders between levels of command. For a more rapid transmission of simple orders, some form of visual or acoustic signaling was normally employed. However, as fronts became wider, weapons became more sophisticated, military administration developed, and logistics tails became longer, the battlefield was no longer under the view of a single commander and better communications were required. Effective command and control required that headquarters had to be informed instantly of events on distant battlefields—more swiftly than was possible by courier.

With the invention of electrical telegraphy and telephony, battlefield communications systems quickly became electronic, which restored timely contact between distant commanders. This contact was only possible by line, however, and it was not until the invention of the radio that commanders were able to conduct limited command and control while on the move.

Although the introduction of electronic communications systems has increased the ability of commanders to communicate over large distances in real-time, there are still quite considerable hurdles to be overcome. Commanders are often seen as having had mobility since the introduction of the radio during World War I, yet large-scale ventures such as brigade attacks were still being undertaken using runners as late as the end of World War II. Most modern armies still struggle to provide timely contact between commanders, particularly when on the move. While all armies strive to achieve real-time situational awareness across the levels of command, the reality is still far from ideal with communications at the lower levels relying on single-channel, shared-access nets that are grossly inadequate to provide the bandwidths required.

The Information Age and the information revolution promise to change all that. Purveyors of Information-Age technologies offer the prospect of ubiquitous battlefield networks through which all battlefield entities are seamlessly integrated so that real-time data can be shared, as and when required. To modern commanders the allure of this promise is irresistible. It has been several hundred years since a commander has had the ability from a convenient hilltop to survey personally the disposition of all friendly and adversary forces. Now, in the Information Age, the modern commander, with senses enhanced by electronic sensors and modern communications systems, can stand on an electronic hilltop and once again "see" whatever portion of the battlefield is desired at whatever detail is appropriate.

This book analyzes the tactical communications systems required to support modern land commanders. (The vulnerability of such systems is addressed in *Electronic Warfare for the Digitized Battlefield,* also published by Artech House.) Here, we develop an architecture to define the tactical communications system required to support future land warfare. The architecture establishes a flexible environment that will support development and acquisition of future battle-space communications systems to support land operations, as well as the modification and enhancement of existing systems to migrate to the target architecture. The architecture is applicable to all modern armies, albeit to varying degrees.

Chapter 1 describes the operational environment of the digitized battlefield and examines the process of command and control that is the core business of the tactical commander. The concept of network-centric warfare is discussed as an example of a doctrine that is emerging to harness the power of the information revolution for application to land warfare.

Chapter 2 provides a brief overview of the technologies that are pertinent to the provision of tactical communications systems. Space restrictions constrain the scope of the descriptions, but the intent is to provide readers with sufficient background to follow the discussion in subsequent chapters.

Chapter 3 provides an overview of army organizational structures, aiming to provide a basic understanding of the size and disposition of army units, army command structures, tactical communications structures, and the roles of support and services. The basic army structure discussed in this chapter applies to most modern armies. The coverage is general; we do not discuss details of any particular army. Many characteristics of the communications system are determined by the structure they support. It is therefore useful to examine the general structure of land forces as a precursor to analyzing the communications systems that support them. Since we are interested in tactical communications systems, we focus on the structure of land forces at the tactical level, that is, at divisional level and below.

Readers familiar with communications technologies and army organizational structures may wish to skip to Chapter 4, which begins by briefly examining the early history of military communications and then focuses on the development of the two major battlefield communications subsystems that are deployed by all modern armies. From this background, a number of verities of tactical battlefield communications are discussed to provide a detailed understanding that can serve as a basis for subsequent analysis of the communications support that can be provided to modern battlefield commanders.

Current tactical communications systems have evolved to meet users' needs as the conduct of warfare has changed, particularly over the last several hundred years. However, if land warfare is to be revolutionized by Information-Age technologies, the tactical communications architecture must become an integral part of a force's ability to prosecute war. This critical interdependency between communications and command and control requires a reconsideration of the architecture of tactical communications systems. Chapter 5 develops an architectural framework to define the tactical communications system. It begins by outlining key design drivers that shape the architecture of a tactical communications system. Options for a mobile tactical communications system are then examined and a suitable framework is developed within which architectural issues can subsequently be considered.

Chapters 6 through 9 then address the major subsystems of the tactical communications architecture. Each subsystem is considered in terms of the fundamental design drivers that direct the provision of tactical communications. Architectural options are analyzed and a preferred option is selected. Consideration is also given to the options available for the migration of legacy subsystems.

Finally, Chapter 10 addresses the critical issue of interfaces. Because there are differing requirements for mobility and capacity, it is not feasible with current or foreseeable technology to provide the tactical communications system as a physically homogeneous network. It is essential, nonetheless, to provide a system that forms a single logical network to facilitate the movement of data throughout the battlespace. This is in line with current trends in commercial networking technology. The interfaces between the different parts of the tactical communications system and between it and other systems play an important part in this integration.

Acknowledgments

We would like to acknowledge the contribution of Bill Blair, Trevor Mahoney, and Dave Rose during early stages of this work.

1

The Need for an Architecture

1.1 Introduction

Success on the battlefield depends on the ability of a commander and staff to process rapidly the huge quantities of information that have been presented to them by the vast array of sensors that may be deployed on the modern battlefield. This information must be accurate and received in a timely manner to allow the preparation of appropriate plans that can then be communicated in a timely manner to those forces that will implement them. Command posts must therefore be connected to their sensors and to subordinates by reliable, survivable communications systems with sufficient capacity. The rapid growth in sensor and weapon systems and the mobility required on the modern battlefield severely tax tactical communications systems to be able to meet the commander's requirements. Yet, at a time when demands on tactical communications systems are increasing dramatically, these systems are rapidly becoming critical to a modern commander's ability to command and control; without them, commanders at all levels are deaf, dumb, and blind.

Traditionally, tactical communications systems comprise two major subsystems. *Trunk communications* exist between major headquarters to provide the backbone of high-capacity circuits required to pass large quantities of information between commanders and staffs. Trunk communications are complemented by the highly mobile and flexible *combat net radio* (CNR) network, which is used for lower-capacity communications, both netted voice and internetted data, from headquarters to combat units and within combat units.

The greater demands of the modern battlefield have a number of effects on communications systems, which must be more flexible and adaptable. No longer can such systems be designed solely to support conventional operations. Modern systems must be modular and expandable to support a wide variety of operations in diverse environments. Additionally, the traditional niches occupied by systems must be expanded. For example, CNR must not only perform its traditional role of voice communications for combat forces, but it must also provide an extension to the voice and data services of the trunk network. Both networks must be seamlessly integrated to allow communication between any two points on the battlefield.

Most modern armies are entering a period of sustained and substantial changes in structure, doctrine, and use of technology, particularly with the current and planned introductions of a wide range of new communications and information systems. In the face of considerable doctrinal and technological change, communications requirements must be reassessed through the development of a comprehensive, flexible architecture for the tactical communications system required to support future land warfare.

1.2 Operational Environment

Arguably the most significant technological revolution in warfare will be in the role of information, and in particular, in the degree of situational awareness made possible by the increasing number of communications and information systems supporting combat forces. Experience in the commercial world has shown that the major lesson for warfare is that conflict in the Information Age will largely be about knowledge and networked organizations will be provided with a major advantage in conflict. A networked battlefield force will be connected via a networked grid that allows situational awareness data to be shared by sensors, command posts, and weapons systems, regardless of whether they belong to the same unit. Information no longer has to be shared along lines dictated by hierarchical communications systems because the network will also facilitate lateral flow. Since the chain of command is no longer restricted to a hierarchical communications system, it can be adjusted quickly to present an order of battle best suited to the task at hand. The adoption of these new technologies, therefore, not only will significantly affect the way armies are commanded and controlled, but also has significant potential to change the way they are organized and trained.

The issue of warfare in the Information Age is addressed in a number of books and articles [1]. For our purposes, we examine tactical communications on the digitized battlefield within the context of the framework articulated by the U.S. Joint Vision 2020 (JV2020) [2]. JV2020 builds upon the conceptual template established in Joint Vision 2010, and has the goal of transforming U.S. forces to create a force that is dominant across the full spectrum of military operations such as conventional warfighting, peace enforcement, peacekeeping, counterterrorism, humanitarian assistance, and civil support. JV2020 also recognizes that the adoption of information technologies is not sufficient to maximize the use of the opportunities made available by the information revolution, and notes that the joint vision can only be realized through a transformation of the necessary doctrine, organization, training, materiel, leadership and education, personnel, and facilities.

Another useful elaboration of the impact of information technology is the emerging concept of *network-centric warfare* (NCW), which is sometimes also called *network-enabled warfare* (NEW). Current warfare could be termed *platform-centric* because the ability to sense and engage a target normally resides on the weapon system ("shooter"). There is therefore only a limited capability for the weapon to engage targets because it can only use the situational awareness generated by its own sensor. If a shooter is able to engage a target that has been sensed by a remote sensor, the passage of weapon data is normally via a direct link that connects the single weapon directly to the single sensor. In NCW, sensors and shooters are connected to a battlefield-wide network through which weapons can engage targets based on a situational awareness that is shared with other platforms. It should be noted that the networking of weapons and sensors does not mean that targets can be engaged randomly; control is still essential to ensure that targets are engaged in accordance with the operational plan.

Alberts, Gartska, and Stein [3] define NCW as "an information superiority-enabled concept of operations that generates increased combat power by networking sensors, decision makers, and shooters to achieve shared situational awareness, increased speed of command, higher tempo of operations, greater lethality, increased survivability, and a degree of self-synchronization." That is, the same combat power can be applied with fewer weapons systems than are currently required.

On the modern battlefield, therefore, the network is a considerable force multiplier. Consequently, tactical communications systems must be ubiquitous across the battlespace and must be fluid, flexible, robust, redundant, and real-time; have integrity and security; have access and capacity; and be joint- and coalition-capable.

Figure 1.1 illustrates the three interlocking grids of NCW (the *information grid*, the *sensor grid*, and the *engagement grid*), and the three major types of participants (*sensors, command elements,* and *shooters*). The information grid provides the infrastructure through which information is received, processed, transported, stored, and protected. The sensor grid contains all sensors, whether they are specialized devices mounted on weapons systems, carried by individual soldiers, or embedded into equipment. The engagement grid consists of all available weapons systems that are tasked to create the necessary battlefield effect. Proponents of NCW envisage that these three grids will exist in space, in the air, on land, and on and under the sea.

The employment of a wireless, nonnodal tactical network allows an army to disperse as required and then mass effects rapidly at an appropriate time and place. Less reliance is required on large command posts, which can be distributed to increase physical survivability without sacrificing processing power. If the philosophy of network-centric warfare is desirable to modern commanders, land forces must be supported by a communications architecture that will allow communication from any one point on the battlefield to any other, as well as from any point on the battlefield to any point in the strategic network. As will be discussed in Chapter 5, these considerations provide significant design drivers for a suitable tactical communications architecture.

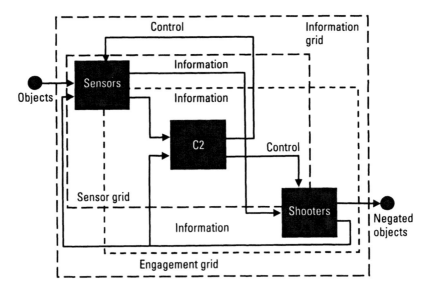

Figure 1.1 The grid arrangement of NCW [4].

1.3 Command and Control

Command and control, and particularly its application in the Information Age, are too broad a subject to be treated in detail here. Interested readers are referred to the many texts that cover the field [5]. We do, however, need to define some terms to ensure a common view of subsequent chapters.

Command is perhaps best described as the authority vested in an individual for the direction, coordination, and control of military forces. *Control* is the means by which command is exercised. The two terms are inextricably linked—there is no point in having the authority to command if there is not an ability to control, and control mechanisms are impotent without the authority to command. Therefore, a commander's business is commonly referred to as *command and control* (C2), which can be described as the process of and means for the exercise of authority of a commander over assigned forces in the accomplishment of the commander's mission. U.S. doctrine adds that command and control functions are performed through an arrangement of personnel, equipment, communications, facilities, and procedures employed by a commander in planning, directing, coordinating, and controlling forces and operations in the accomplishment of the mission [6].

1.3.1 The C2 Cycle

The processes underlying C2 are illustrated by the *C2 cycle* shown in Figure 1.2. Although a very simple model, the C2 cycle is useful as a framework for the application of command and control at any level, although it should be applied cautiously [7]. For our purposes, it is also useful to visualize the impact that communications systems have on the modern battlefield.

The C2 cycle is also called the *decision cycle*, the *OODA (OUDA) loop* (for the elements of *o*bservation, *o*rientation (*u*nderstanding), *d*ecision and *a*ction), or the *Boyd cycle* [8] after the retired U.S. Air Force Colonel John Boyd who pioneered the concept.

The cycle is continuous, but it can be considered to start with the process by which data is gathered by a wide range of *surveillance and target acquisition* (STA) devices deployed to act as the commander's eyes and ears. Surveillance data from modern sensors is invariably collected and reported in digital form, and it is the rapid increase in the number of sensors and surveillance systems that is predominantly responsible for the explosion in digital transmission requirements on the modern battlefield. Surveillance data can only reach the commander if effective, survivable communications systems

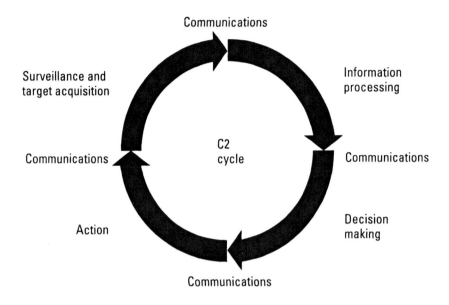

Figure 1.2 The C2 cycle.

are available from the sensor system through to the data processing facilities in the command post. The commander only has eyes and ears if robust, flexible communications systems with sufficient capacity can be provided.

In the next phase, *information processing*, the raw sensor data is processed, fused, and analyzed so that it can be turned into information. The commander and staff could not cope with the volume of raw sensor data coming into the headquarters. Not only would the sheer volume of it be overwhelming, but also the raw data is unintelligible to humans and must be filtered and displayed in an appropriate format before the commander and staff can take action on it. Automation of this process is essential if it is to be completed in a timely manner.

Once the sensor information has been processed into a view that provides the commander with the right information in the right format at the right resolution in sufficient time to allow the conduct of an adequate appreciation, or estimate, of the situation. In the *decision-making* phase, the commander makes a number of decisions and finalizes a plan, following which orders are conveyed to subordinate units through data and voice networks, whereupon *action* is taken on the commander's plan. Few plans last beyond first contact with the enemy, however, so the C2 cycle must continue and information begins to flow to allow the commander to control the action of subordinate units and to task STA assets to monitor operations.

There are many more complex models for command and control [9], but the C2 cycle is adequate for our purposes here since it is evident even from the simple model that a commander's success is heavily reliant on reliable communications systems. We should further note the heavy reliance that communications systems have on the electromagnetic spectrum.

1.4 An Increased Vulnerability

While the Information Age has produced a revolution in military operations that provides great promise of decisive advantage on the modern battlefield to the commander who can gather and exploit information most effectively, there is a significant dark side to the information revolution. Although Information-Age technologies offer the promise of enormous increases in command capability, they also bring with them a heavy dependence on the electromagnetic spectrum, which has the potential to create a significant vulnerability that may offset the advantages offered.

If these Information-Age systems are destroyed, degraded, or deceived, the commander and staff are unable to prosecute war adequately. Without communications on the modern battlefield the commander is increasingly unable to command. Domination of the electromagnetic spectrum is therefore a crucial component of most modern military operations and the capability to conduct electronic combat and dominate the electromagnetic spectrum is now a recognized component of any modern force structure.

Electronic warfare (EW) can be defined as "the use of the electromagnetic spectrum to degrade or destroy an adversary's combat capability (including degrading or preventing use of the electromagnetic spectrum as well as degrading the performance of adversary equipment, personnel and facilities); or to protect friendly combat capability (including protecting friendly use of the electromagnetic spectrum as well as friendly equipment, personnel and facilities that may be vulnerable to attack via the electromagnetic spectrum)" [10]. Figure 1.3 illustrates how EW pervades all aspects of the modern battlefield and has the potential to impact on all elements of the C2 cycle.

It is essential that the development of tactical communications systems take into account the extent to which the electromagnetic spectrum must be dominated on the modern battlefield.

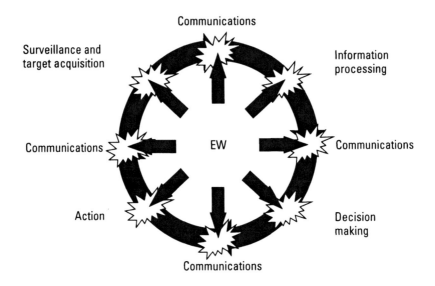

Figure 1.3 The potential impact of EW on the C2 cycle.

1.5   Tactical Communications for the Digitized Battlefield

This book analyzes the *tactical communications* required to support operations on the *digitized battlefield.*

1.5.1   Tactical Communications

The term *tactical* has been chosen to represent the level at which communications are considered. Three broad levels of warfare are defined [6].

Strategic level of war.   The level of war at which a nation, often as a member of a group of nations, determines national or multinational (alliance or coalition) security objectives and guidance, and develops and uses national resources to accomplish these objectives. Activities at this level establish national and multinational military objectives, sequence initiatives, define limits and assess risks for the use of military and other instruments of national power, develop global plans or theater war plans to achieve these objectives, and provide military forces and other capabilities in accordance with strategic plans.

Operational level of war.   The level of war at which campaigns and major operations are planned, conducted, and sustained to accomplish strategic objectives within theaters or areas of operations. Activities at this level link tactics and strategy by establishing operational objectives needed to accomplish the strategic objectives, sequencing events to achieve the operational objectives, initiating actions, and applying resources to bring about and sustain these events. These activities imply a broader dimension of time or space than do tactics; they ensure the logistic and administrative support of tactical forces, and provide the means by which tactical successes are exploited to achieve strategic objectives.

Tactical level of war. The level of war at which battles and engagements are planned and executed to accomplish military objectives assigned to tactical units or task forces. Activities at this level focus on the ordered arrangement and maneuver of combat elements in relation to each other and to the enemy to achieve combat objectives.

The levels of war are rarely distinct and often overlap, particularly in operations other than war. In this book we use the term *tactical* to refer to operations at divisional level and below. In terms of the provision of communications, this level is the most difficult.

Military communications systems are not unique in this regard; in fact, they are an extreme example of the difficulty faced in commercial systems when attempting to deliver high-capacity communications to end-users. Large bandwidths are relatively easy to obtain in trunk links, particularly when fiber-optic cable can be used between cities and exchanges. Distributing large amounts of information below exchanges is problematic, however, particularly when the users are mobile. In the commercial environment, the links between users and the exchanges are generically called the "last mile" and represents the major stumbling block to the distribution of large-bandwidth services such as multimedia applications.

As we describe in Chapter 5, the "last tactical mile" is also the major difficulty in military communications. It is relatively straightforward to provide high-capacity communications at the strategic and operational levels; in fact, commercial systems can invariably be used with only limited modification. In the tactical environment, and in particular in the land tactical environment, there are few commercial solutions that are directly applicable. It is this difficult area of land tactical communications systems that is the subject of this book.

1.5.2 The Digitized Battlefield

The terms *digitized battlefield* and *battlefield digitization* have been adopted to refer to the automation, through digital networks and processes, of command and control operations across the full breadth of the battlespace. This integration of ground, air, and space nodes (sensors, communications, command and weapons nodes) into a seamless digital network requires the fully compatible digital exchange of data and common operating pictures to all nodes. Security, compatibility, and interoperability factors dominate the drive toward full digitization across the entire battlespace.

The notion of battlespace is very important, as joint operations are essential. However, the difficult issues in tactical communications are not ones of joint operations; rather they are in solving the difficult issues of high-capacity communications across the last tactical mile. The word *battle-field* has therefore been retained here to emphasize the importance of the terrestrial component of battlespace operations in driving the architecture of a land tactical communications system. The joint environment is quite obviously important, however, and we consider battlespace communications as one of the many interfaces required for a battlefield communications system.

1.6   The Need for an Architecture

Land forces must also be supported by an organic, field-deployable tactical communications system that meets minimum essential requirements for communications to support command and control. This tactical communications system is an organic asset that is part of the force's combat power through which the essential communications requirements of the force can be guaranteed in any deployment. This also requires that the network is modular so that units and subunits are self-supporting when deployed separately from the main force and still retain communications functionality. These requirements cannot be supported by a global, or even battlespace–wide, network [11] and can only be supported within a tactical communications system.

Most armies currently field trunk and CNR systems that were conceived some 15 to 20 years ago in a very different operational and technological environment. While users and project staff have ensured that projects have kept pace with changes in environment and technology, both systems will need considerable modification and enhancement to provide robust tactical communications to support future land warfare.

As described earlier, modern armed forces must be able to defeat adversaries across a wide range of operations such as conventional warfighting, peace enforcement, peacekeeping, counterterrorism, humanitarian assistance, and civil support. Each of these types of operations requires a different type of deployment. Most small- to medium-sized armies cannot afford to maintain a range of different types of force structure, each capable of conducting one or more of the above operations. Rather, land forces must be flexible and be equipped and trained to cope with a wide range of deployments. The tactical communications system must also be equally able to support land forces in any of their roles. As is demonstrated throughout this book, such flexibility cannot be provided adequately by existing tactical communications systems or by existing commercial systems.

Most armies are already planning projects to replace or enhance current CNR and trunk capabilities. In addition, new technologies such as portable satellite communications and theater broadcast offer novel capabilities to support battlefield operations. To varying degrees, each of these projects is addressing the changes discussed earlier. However, they are tending to do so based on the traditional delineation between the different tactical communications systems. It is therefore unlikely that an optimum solution will be achieved without the development of an overarching communications architecture that provides a framework within which communication projects can progress. Management of rapidly evolving technology in the capability-development and acquisition processes requires a corporate approach if integrated and cohesive solutions are to be achieved. Currently, the capability-development process tends to be driven from the bottom up. Creation and maintenance of the information edge require a top-down approach (albeit with some bottom-up elements) where broad guidelines and policies for the operation and design of systems will mandate functional interoperability, upgrade and migration plans, and common equipment standards.

Therefore, defense forces cannot afford to continue to develop or modify battlefield communications systems on current paths based on traditional directions. An appreciation must be conducted into the communications support required to support the required spectrum of operations, with the aim of developing an architecture [12] that provides the foundation for the seamless flow of information and interoperability among all tactical communications systems operating in the battlespace to support land operations.

In the remainder of this book we develop such an architecture to define the tactical communications system required to support future land warfare. The architecture establishes a flexible environment that will support development and acquisition of future battlefield communications systems to support land operations, as well as the modification and enhancement of existing systems to migrate to the target architecture. The architecture is applicable to all modern armies, albeit to varying degrees.

The architecture developed here broadly conforms to the *C4ISR Architecture Framework* [13]. Chapters 3 and 4 provide an operational architecture view by describing the concepts of operations and force structure that drive the need for tactical communications systems. The system architecture view is provided in Chapter 5, which describes an overview of the system architecture, whose components are described further in Chapters 6 through 10. These later chapters also provide a preliminary view of the technical architecture through an introduction to the technical standards and protocols that guide current and future implementations.

Endnotes

[1] Further suggested reading on warfare in the Information Age:

Adams, J., *The Next World War*, London: Random House, 1998.

Alexander, J., *Future War: Non-Lethal Weapons in Twenty-First Century Warfare*, New York: St. Martin's Press, 1999.

Allard, C., *Command, Control, and The Common Defense*, New Haven, CT: Yale University Press, 1990.

Arquilla, J., and D. Ronfeldt, (eds.), *In Athena's Camp: Preparing for Conflict in the Information Age*, Santa Monica, CA: RAND, 1997.

Bellamy, C., *The Future of Land Warfare*, New York: St. Martin's Press, 1987.

Campen, A., and D. Dearth, *CyberWar 2.0: Myths and Reality*, Fairfax, VA: AFCEA International Press, 1998.

De Landa, M., *War in the Age of Intelligent Machines*, New York: Zone Books, 1991.

Gordon, A., *The Rules of the Game: Jutland and British Naval Command*, Annapolis, MD: Naval Institute Press, 1996.

The International Institute for Strategic Studies (IIS), *Strategic Survey 1995–1996*, London, U.K.: Oxford University Press, 1996, p. 30.

Leonhard, R., *The Principles of War for the Information Age*, Novato, CA: Presidio Press, 1998.

Macgregor, D., *Breaking the Phalanx: A New Design for Landpower in the 21st Century*, Westport, CT: Praeger Publishers, 1997.

Peters, R., *Fighting for the Future: Will America Triumph?* Mechanicsburg, PA: Stackpole Books, 1999.

Pfaltzgraff, R., and R. Shultz, (eds.), *War in the Information Age: New Challenges for US Security*, Washington, D.C.: Brassey's, 1997.

Rooney, D., V. Kallmeier, and G. Stevens, *Mission Command and Battlefield Digitization: Human Sciences Considerations*, DERA Report DERA/CHS/HS3/CR980097/1.0, March 1998.

Scales, R., *Future Warfare*, Carlisle, PA: U.S. Army War College, 1999.

Toffler, A., and H. Toffler, *War and Anti-War: Survival at the Dawn of the 21st Century*, Boston, MA: Little, Brown and Company, 1993.

Van Trees, H., "C3 Systems Research: A Decade of Progress," in *Science of Command and Control: Coping with Complexity*, S. E. Johnson and A. H. Levis (eds.), Fairfax, VA: AFCEA International Press, 1989.

Waltz, E., *Information Warfare Principles and Operation*, Norwood, MA: Artech House, 1998.

[2] "Joint Vision 2020," Director of Strategic Plans and Policy, J5 Strategic Division, Washington, D.C.: U.S. Government Printing Office, June 2000.

[3] Alberts, D., J. Gartska, and F. Stein, *Network Centric Warfare*, CCRP Publication Series, Washington, D.C.: U.S. Department of Defense, 1999.

[4] Cebrowski, A., and J. Garstka, "Network-Centric Warfare: Its Origins and Future," *Naval Institute Proceedings*, 1997.

[5] Further reading on command and control can be found in:

Dupuy, T., *Understanding War: History and Theory of Combat*, New York: Paragon House Publishers, 1987.

Echevarra, A., "Tomorrow's Army: The Challenge of Nonlinear Change," *Parameters*, Autumn 1998, p. 11.

van Creveld, M., *Command in War*, Cambridge, MA: Harvard University Press, 1985.

van Creveld, M., *The Transformation of War*, New York: Free Press, 1991.

[6] Joint Publication 1-02, "Department of Defense Dictionary of Military and Associated Terms," Washington, D.C.: Office of the Joint Chiefs of Staff, 1994 (as amended Sept. 2000).

[7] Kallmeier, V., et al., "Towards Better Knowledge: A Fusion of Information Technology, and Human Aspects of Command and Control," *Journal of Battlefield Technology*, Vol. 4, No. 1, March 2001, pp. 34–43.

[8] For a good description of the Boyd cycle, see: Lind, W., *The Maneuver Warfare Handbook*, London, U.K.: Westview, 1985, p. 5.

[9] See in particular: Boyes, J., and S. Andriole, (eds.), *Principles of Command and Control*, Washington, D.C.: AFCEA International Press, 1987; and Andriole, S., (ed.), *Technology for Command and Control Systems Engineering*, Fairfax, VA: AFCEA International Press, 1990.

[10] Further information on communications EW can be found in: Frater, M., and M. Ryan, *Electronic Warfare for the Digitized Battlefield*, Norwood, MA: Artech House, 2001.

[11] The recent RAND report—Libicki, M., *Who Runs What in the Global Information Grid: Ways to Share Local and Global Responsibility*, Rand Report, MR-1247-AF,

2000—makes interesting reading and concludes that it is generally undesirable for the provision of operational information to be centralized.

[12] *Architecture* is defined as "the organizational structure of a system or component." See IEEE 610.12-1990, *IEEE Standard Glossary of Software Engineering Terminology*, New York: The Institute of Electrical and Electronics Engineers, 1990.

[13] C4ISR Architectures Working Group, *C4ISR Architecture Framework*, Version 2.0, Washington, D.C.: U.S. Department of Defense, Dec. 18, 1997.

2

An Introduction to Communications Technology

2.1 Introduction

Communication can be defined in many ways. For our purposes we will define it simply as *the exchange of information*, which, for humans, is normally in one of three forms: verbal (spoken words), written (text), or visual (images or sequences of images such as video). Every day, most of us will use each of these forms. Often, however, the recipient of the information is beyond the range of normal human visual or oral/aural processes. In that case, we must exchange information with the aid of a *communications system.*

Early long-distance communications systems employed messengers or runners, or utilized some form of visual or acoustic signaling technique. We do not address these systems here (we discuss them in more detail in Chapter 4), except to note that courier systems still play a large role in modern communications, where they can often be the most cost-effective technique for transferring bulky information. Visual and acoustic techniques relied on drums, call posts, cannon shot, fire beacons, flag drill, heliograph, and optical telegraphy, to name but a few methods. Although some techniques linger (most modern navies still use flag and light signals for short-range communications), these forms of signaling were quickly replaced by electrical telegraphy in the mid-1800s. Since then, the modern forms of voice, data, and video communications have evolved, all of which rely on the electrical signal as the means of communication.

This chapter provides a brief overview of the communications technology that is used to convey information through the transmission of electrical signals.

2.2 Communications Model

Figure 2.1 [1] illustrates the major functions that are performed as part of a modern digital communications system [2]. The *source* converts the real-world original signal into an electrical signal, which is increasingly in a digital form. This signal is then compressed through *source coding* to remove any redundancy to ensure that the signal occupies the least amount of the channel bandwidth. *Encryption* then ensures that the signal is encoded for privacy, and *channel coding* prepares the signal for the noise and distortion that will occur in the channel. A number of signals are then *multiplexed* together and then translated in frequency (*modulated*) up to a frequency that is appropriate for the particular channel in use. *Frequency spreading* is used to reduce the signature of the transmission, and *multiple-access* techniques are necessary to allow a number of users to share a channel. These functions are described in more detail in the remaining sections of this chapter.

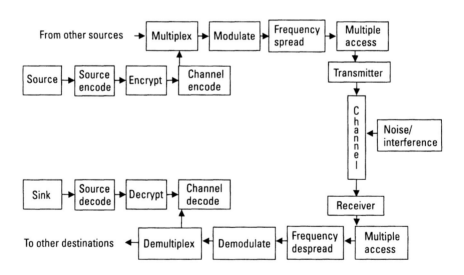

Figure 2.1 Block diagram of a typical modern digital communications system [1].

2.3 Source/Sink

A communications system passes real-world information from the source to the sink. The source converts the information into an electrical signal that is passed across the communications channel and is finally converted by the sink back into a form of information that can be digested by the recipient.

Electrical signals can be transmitted in a wide range of forms—in this section we are concerned with the two broad forms of *analog* and *digital* signals. Whatever their form, signals used in communications systems are often complex and their analysis would be very difficult in their complete form. Fortunately, however, all signals can be considered to be a combination of simple basic building blocks (sine waves) that can be analyzed and from which information about the more complicated signal can be inferred.

2.3.1 Analog Signals

The *sinusoidal function* (or *sinuate* or *sinusoid* or *sine wave*) is shown in Figure 2.2.

2.3.1.1 Frequency, Wavelength, and Phase

From the sinusoidal waveform we can determine a number of basic properties. It is symmetrical, varying between magnitudes of +A and −A. It is regular or periodic, that is, the waveform repeats itself in a series of cycles. Figure 2.2 shows the extent of one cycle or period, *T*, which is measured in seconds. The period of the waveform is perhaps best simply understood as the amount of time after which the waveform begins to repeat itself. Another important

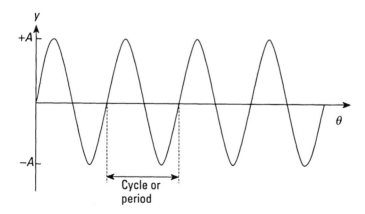

Figure 2.2 A simple sinusoid of the form $y = A \sin(\theta)$.

property defined for the wave is the frequency, that is, how many cycles (or periods) of the wave occur per second. The frequency, f, is therefore defined as $f = 1/T$ and is measured in s^{-1} or hertz (Hz). The frequency of the waveform is defined by how many cycles there are per second. A high frequency wave therefore has more cycles per second than a low frequency wave.

In the time of one period, the wave travels one wavelength, λ, which is measured in meters and is the distance between points of similar amplitude in the propagating waveform. The wavelength of a sine wave is related to its period by the velocity of propagation of the wave, v_p. That is, $v_p = \lambda T$, or $\lambda = v_p T$. The velocity of propagation of an electromagnetic wave in the atmosphere is the speed of light, c, which equals 3×10^8 ms^{-1}. The wavelength (distance) of the wave is therefore related to the period (time) of the wave by the relationship: $c = \lambda/T$ or $\lambda = cT$. Now since $T = 1/f$, $\lambda = c/f$ or $f = c/\lambda$.

At the beginning of this section we noted that a complex waveform can be expressed as the sum of a number of sine waves at different frequencies. An example of this is demonstrated in Figure 2.3, where a complex wave is expressed as the sum of 4-, 5-, and 6-kHz waves.

Two sine waves can have identical amplitudes and frequencies, but, as illustrated in Figure 2.4, if one is delayed compared to the other, they are identifiably different. This difference is called phase, which is the third important property of a sine wave. In communications systems we need to understand the phase of received signals because in-phase signals add to produce a stronger signal, whereas out-of-phase signals tend to cancel each other.

2.3.1.2 Frequency Domain

It is often useful to view the waveform in the frequency domain where the frequency is plotted against the amplitude of the signal. This is illustrated in

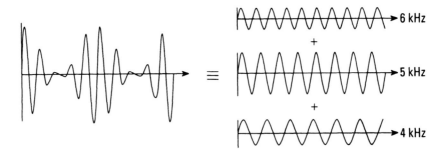

Figure 2.3 A complex waveform expressed as a sum of sine waves.

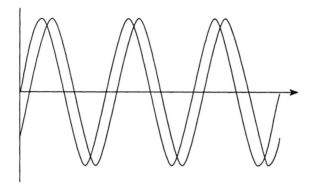

Figure 2.4 Two sine waves with identical amplitudes and frequencies but different phases.

Figure 2.5, where a sine wave with a period of 1 ms (and therefore a frequency of 1 kHz) is shown in Figure 2.5(a), the time domain, and Figure 2.5(b), the frequency domain. In the frequency domain a single frequency sine wave is depicted by a frequency spike of amplitude V_m located at 1 kHz on the frequency axis.

This frequency-domain view is very useful for the analysis of signals because it is difficult to represent complex signals as combinations of sine waves in the time domain. Yet in the frequency domain, the representation consists simply of a spike of the appropriate amplitude at the appropriate frequency for each component. For example, the complex waveform in Figure 2.6 can now be represented more simply in the frequency domain.

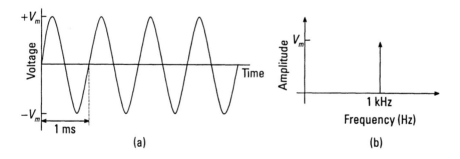

Figure 2.5 A 1-kHz sine wave in (a) the time domain, and (b) the frequency domain.

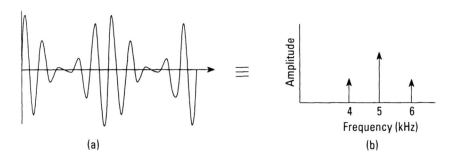

Figure 2.6 A complex waveform in (a) the time domain, and (b) the frequency domain.

Bandwidth. The frequency-domain representation also allows us to simplify the view of a waveform as depicted in Figure 2.7(a) by an equivalent representation as shown in Figure 2.7(b). The collective representation is useful when we want to consider the range of frequencies as a band and not necessarily bother with each individual frequency. The band is defined by the upper and lower limits of the frequencies present and no attempt is made to define the amplitudes of the constituent frequencies, either collectively or individually. The frequency-domain representation of a range of frequencies leads us conveniently to the measure of bandwidth, which is best considered to be the difference between the highest frequencies that have to be passed by the system.

2.3.1.3 Speech Signals

Humans can produce sounds between 100 and 10,000 Hz and the human ear can hear sounds in the range of 15 to 15,000 Hz [known as audio frequencies (AF)]. Most of the energy for speech signals is contained in the lower frequencies, however, and most useful frequencies are contained

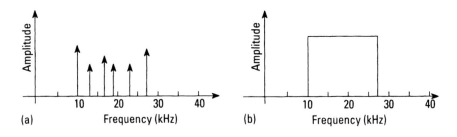

Figure 2.7 A number of sine waves shown (a) individually, and (b) collectively as a band.

between about 300 Hz and 3,400 Hz. As illustrated in Figure 2.8, the bandwidth of speech signals is therefore normally considered to be 3.1 kHz.

2.3.1.4 The Electromagnetic Spectrum

Due to the size of the electromagnetic spectrum, it is convenient to break it up into sections, which exhibit common properties. We are interested in those portions of the spectrum that can be used for communications systems, notably those frequencies contained in the radio frequency (RF) ranges. In these regions of the spectrum, the International Telecommunication Union (ITU) Radio Regulations define the bands of frequencies outlined in Table 2.1 (note that the RF range also overlaps with the AF range of 15–15,000 Hz).

Extremely low frequency (ELF). The ELF band is of little use in communications due to the extremely small bandwidth available and the enormous antennas required. Even these large antennas are inefficient, however, and radiated powers are small even when very large transmitter powers are used. Despite these disadvantages, however, ELF has one redeeming feature that makes its use a viable proposition for communications with submarines. ELF signals suffer very little attenuation by any medium, particularly the atmosphere and seawater. Propagation is primarily via surface wave, and the signal is very stable with little diurnal or seasonal variation. In addition to submarine communications, some of the audible frequencies are contained in this band. Although not part of communications frequencies, electric power is generated in this band (60 Hz in the United States).

Voice frequency (VF). VF frequencies are those generated by the human voice and are therefore critical to communications systems as they will often

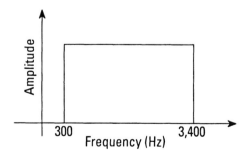

Figure 2.8 The range of useful speech frequencies.

Table 2.1
RF and AF Portions of the Electromagnetic Spectrum

Frequency	Designation	Wavelength
30–300 Hz	ELF	10–1 mm
300–3,000 Hz	VF	1–0.1 mm
3–30 kHz	VLF	100–10 km
30–300 kHz	LF	10–1 km
300–3,000 kHz	MF	1–0.1 km
3–30 MHz	HF	100–10m
30–300 MHz	VHF	10–1m
300–3,000 MHz	UHF	1–0.1m
3–30 GHz	SHF	100–10 mm
30–300 GHz	EHF	10–1 mm

be the required input and output of a system. VF frequencies do not propagate very far so the communications system will not pass VF directly but will need to transpose the band up to a higher frequency range so that it can be passed across the channel.

Very low frequency (VLF) and low frequency (LF). VLF and LF suffer from the same problems as ELF in that they have small bandwidths and antennas are large and inefficient, leading to low radiated powers. However, both bands have similar advantages as ELF and have low attenuation through the atmosphere and seawater. VLF and LF are therefore used for some submarine communications and maritime radio navigation systems. These bands have more use for communications than ELF, since they have a larger bandwidth. Propagation is primarily by surface wave and has little seasonal or diurnal variation. However, communications are prone to static and to interference from other radio frequencies.

Medium frequency (MF). The lower portion of the MF band is useful for communications systems that require reasonably stable transmission over moderately long distances. Antenna sizes are reasonable, but substantial transmitted powers are required for reliable communications since appreciable atmospheric noise occurs in this band. Propagation is predominantly by surface wave, although there is a sky-wave component above 2 MHz. This band is commonly used for commercial radio broadcasting and is also used for fixed services, maritime mobile service, maritime and aeronautical navigation, and amateur communication.

High frequency (HF). The HF band provides relatively reliable propagation over long distances. Reasonably efficient antennas can be built requiring low radiated power. Propagation is principally by sky wave and therefore depends on the vagaries of the ionosphere, with fading and multipath effects often limiting the speed of communication. Surface wave communications over short distances (~50 km) are also possible. The band is used for fixed services, mobile services, amateur transmissions, broadcasting, and maritime mobile service. Until the advent of satellite communications, the HF band provided the primary means of long-distance radio communication.

Very high frequency (VHF) and ultra high frequency (UHF). Both of these bands can be used for communications systems, such as television, which require the transmission of large bandwidths over short distances (generally line-of-sight). Small, directional antennas are economical and effective. If powerful transmitters and high-gain antennas are used, reliable long-distance propagation is possible in the UHF band by using waves scattered by turbulence in the troposphere. The VHF and UHF bands are used for fixed communications services such as radio relay, ground-to-air communications, mobile services, and television. Other noncommunications services include radar, space research, radio astronomy, and telemetry.

Super high frequency (SHF) and extremely high frequency (EHF). SHF and EHF are known as the microwave bands and their wavelengths are short enough to be propagated by highly directional antennas and waveguides. Propagation ranges are limited to line-of-sight, but long-distance communications can be achieved by employing a series of radio-relay stations. These portions of the spectrum are used for television and high-speed data services requiring large bandwidths.

Communications frequencies other than RF and AF. The majority of the frequencies used for communications are contained within the RF portion of the electromagnetic spectrum. Optical frequencies are also used for communications, either in free-space laser communications or, more commonly, confined to the optical fiber waveguide. More recently, infrared communications techniques are also employed to provide the transmission medium for wireless local area networks (WLANs).

2.3.2 Digital Signals

Many sources produce analog signals, since the world is naturally a continuous, or analog, medium. Many modern sources, however, produce digital signals since the message is originated by electronic equipment such as a computer. Additionally, analog information is also often converted into a digital form since the transmission of digital signals has a number of significant advantages over analog signals, namely:

- All digital signals have the same form (i.e., they are either "ones" or "zeros"), easier storage, and easier multiplexing, switching, and redirection.
- Digital signals have lower susceptibility to noise and are more easily regenerated.
- If a communications system is all-digital, only one type of circuitry is required.
- Security equipment is more easily implemented.
- Error correction and detection can be more easily implemented.

Digital signals do have some disadvantages in that greater bandwidth is sometimes needed (which may be prohibitive, especially at HF) and the signals are more complex. The advantages, however, significantly outweigh the disadvantages.

Digital signals imply some coding or quantization of an original signal to a limited number of discrete values or levels each of which can be identified by a digit as one element of the code. The codes most commonly used are binary codes and the element of the code is called a binary digit (bit). As an element of a binary code, the bit can take one of two values: 0 or 1. Voltage levels, such as ±15V, represent the states of the symbol, as illustrated in Figure 2.9.

There are two ways to transmit digital signals from source to sink. In the baseband transmission of digital signals, voltage levels are transmitted directly. Alternatively, the digital signal can be used to modulate an analog

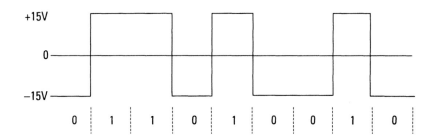

Figure 2.9 An example of a digital signal.

carrier for transmission over longer distances. For the moment, consider digital signals transmitted in baseband.

2.3.3   Baseband Digital Signals

Analog signals suffer as they pass along the communications channel because they are attenuated (reduced in size) and have their amplitudes modified by the addition of unwanted signals in the form of noise. At various points along the channel, attenuation of the signals can be compensated for by amplification of the signal, although the unwanted noise will also be amplified. Therefore, amplifiers increase signal strength at the receiver but also increase the level of received noise and distortion.

So why are digital signals less susceptible to noise? Surely the digital signal is attenuated in the same way as analog signals as it passes along the channel. The answer lies in the fact that the attenuation and noise in the channel do have the same effect on the signal, but not the same effect on the information contained within the signal. In analog signals the information is contained in the small variations in the signal with time. Attenuation in the channel and the addition of noise tend to distort the signal so that some of the original information is lost, often irretrievably. Attenuation can be addressed by amplification at the sink. As we noted earlier, however, amplification of a weak, noisy signal will tend to produce a stronger, noisier signal.

The information in digital signals is contained in the gross value of a bit at any point of time, not in the small variations of the signal. As the bit is attenuated by the channel and noise is added, the information content remains unchanged until the bit can no longer be distinguished as its original value. Therefore, as illustrated in Figure 2.10, as long as a "1" can be distinguished from a "0," the information has not been lost and the signal can be regenerated and received perfectly at the sink. For digital signals the

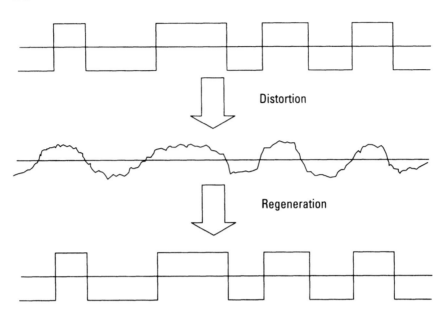

Figure 2.10 Regeneration of a digital signal.

equivalent to analog amplification is regeneration and digital signals can be regenerated as often as required, without the unwanted amplification of noise. After each regeneration, the digital signal is a precise copy of the original.

2.3.4   Information Rate

In addition to defining the rate of transmission of the data or symbols, it is also important to examine the information rate, or the rate at which information is transferred from the source to the sink, because this is most often not the same as the signaling rate.

For example, assume a simple, hypothetical data frame in which there are 100 bits to be transmitted at, for example, a transmission rate of 50 bps, which will take 2 seconds. But only 76 bits of the frame contain information; the remaining bits are required to administer the transmission link. Since this administrative data is essential but does not contain any information, it is called *overhead*. In the 2 seconds that it takes to transmit the 100 bits of data, only 76 bits of information will be sent, so that the *information rate* is only 38 bps (even though the transmission rate is 50 bps).

2.3.5 Asynchronous/Synchronous Transmission

Data transmission can be either *asynchronous* or *synchronous*. Asynchronous transmission occurs without significant prior coordination between the source and the sink. Each block of data is sent by the source, which receives warning of each block by a start bit and notification of the end of the block with a stop bit. Identification of the end of the block is essential since there may be a significant idle period until the next block arrives. Asynchronous transmission has advantages on some links, but for high-speed links, the time spent in coordination is generally prohibitive.

Synchronous transmission relies on both the source and sink running oscillators that are continually synchronized so that only a few coordinating (synchronization) pulses are contained in each block of data. Generally synchronous transmission for direct connections is used for data rates in excess of 1,200 bps.

2.4 Source Coding

Source coding involves the transformation of the information source into some efficient description that reduces the amount of bandwidth required to transmit the source data across the communications system [3].

In *pulse-code modulation* (PCM) the analog signals are first sampled to produce pulses whose amplitudes are proportional to the amplitude of the analog waveform at the time at which the sample was taken. The pulses are then encoded into a binary code that is transmitted as a digital stream. At the receiver these PCM codes are decoded into pulses that are then used to reconstruct the analog waveform. Digitized voice with a maximum frequency of 3.4 kHz is usually sampled at 8 kHz. If an 8-bit coder is used (i.e., 256 quantizing levels), then the transmission rate of the channel is $8,000 \times 8 = 64$ Kbps.

Delta modulation can be considered to be an incremental PCM system where, rather than transmitting the absolute amplitude at each sampling, only the changes in signal amplitude from sampling instant to sampling instant are transmitted. This provides a simpler system and requires less information to be transmitted. The range of signal amplitudes is divided into a similar number of quantization levels as in PCM systems. However, at each sampling, the presence or absence of only one transmitted pulse contains the information. The delta-modulation code is therefore a one-element code in which a "1" is transmitted if the sampled signal at any sampling instant is greater than the immediately previous sampling instant. A "0" is transmitted if the sampled signal is lower. The delta modulator only requires 1 bit per sample. However, to avoid slope overload, it is normal to sample at twice the Nyquist rate (i.e., for speech: 16,000 samples per second) so that the bit rate required to transmit delta-modulated signals is only $1 \times 16,000 = 16$ Kbps. Some modern systems operate as low as 9.6 Kbps.

Comparison of PCM and delta modulation.   Delta modulation has a number of advantages over PCM: lower bandwidth required; better resistance to noise since an error in one bit is not serious, while an error in the most significant bit of a PCM code will make a large difference to the decoded value; easier synchronization; and simpler equipment. Delta modulation has a couple of disadvantages, however, in that it leads to a signal of poorer quality and it has increased error due to slope overload. PCM is normally used in commercial telephone networks where high quality is desired. Delta modulation has found wide application in military communications where low-bit-rate systems are preferred at the expense of some quality.

Vocoders.   A PCM or DM coder does not know anything about the signal being coded and simply follows the signal as it is presented and codes it as a digital bit stream. When transmission bandwidth is limited, such as at HF, more bandwidth-efficient coding techniques can be provided by *vocoding* techniques. A vocoder does not blindly follow the input waveform but uses some knowledge of the source's characteristics to create a model of the source (voice) waveform and send some form of description of the model's parameters to the receiver. The vocoder analyzes the properties of speech (excitation, formants, pitch, and volume) and estimates the dominant characteristics and transmits them to the receiver, which uses them to recreate the source waveform and to synthesize the voice output. Vocoders require very low data rates (2.4 Kbps is common in military systems) but produce lower quality output, usually sacrificing speaker recognition.

Adaptive prediction.   Many analog signals are not best digitized using uniform quantization steps. For example, in speech small signal amplitudes are more common than large ones. For these signals the quantization noise can be reduced by using smaller steps for lower amplitudes and larger steps for higher amplitudes. Adaptive prediction techniques include *adaptive differential pulse-code modulation* (ADPCM), used to provide 32-Kbps telephone-quality speech; *continuously variable slope delta modulation* (CVSDM), commonly used to provide 16-Kbps channels in military networks; and *linear predictive coding* (LPC), which forms the basis of the *regular pulse excited*

(RPE) and *codebook-excited linear predictive* (CELP) coders used in cellular telephone networks.

Vector quantization. The coders discussed so far are scalar in that they form a single output sample for every input sample. Vector quantizers form a single output sample for a vector (or block) of input samples and can therefore achieve very large coding gains.

Transform coding. In transform coding, an invertible transform is applied to a block of original data and the transform coefficients are quantized and then transmitted. The decoder uses the quantized coefficients to invert the transform, and a distorted version of the original data is recovered. The amount of distortion in the received data is controlled by the degree of quantization applied at the encoder. Common transforms include the discrete Fourier transform (DFT), discrete cosine transform (DCT), discrete Walsh-Hadamard transform (DWHT), and the discrete Karhunen-Loeve transform (DKLT).

Entropy coders. Techniques such as block and transform coding rely on the presence of structural redundancy in the source data. These techniques can also be supplemented by the efficient removal of statistical redundancy from digital data, through the selection of an efficient binary representation of the source data. For example, in a simple system, commonly occurring symbols in the source data could be replaced by short code words, while symbols that occur less frequently could be replaced by long code words, thereby reducing the entropy of the source. Examples of this type of entropy coder are the Huffman code, arithmetic codes, Lempel-Ziv codes, and the run-length code (RLC).

2.5 Encryption

An important advantage of digitization, particularly for military communications, is that the signals can then be encrypted. With analog signals, only limited security can be implemented through the use of scramblers. However, sophisticated coding can be achieved through the use of *ciphers* or *codes.*

Encryption protects digital data by transforming the original data (*plaintext*) into a different form (*ciphertext*) that can be revealed without disclosing the original data [4]. The basic structure of a secure communications system employing encryption is shown in Figure 2.11. Encoding (known as

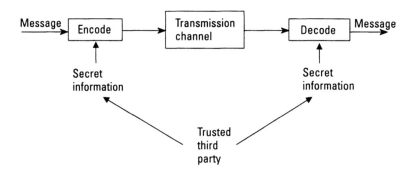

Figure 2.11 Generic structure of a secure communications system.

encryption) and decoding (decryption) involve a security-related transformation. Secret information, which is distributed by a trusted third party, is usually used to modify this transformation so that the security of the system can be maintained even if the encoding and decoding algorithms are widely known. This secret information is often referred to as a *shared secret* or *key*.

In symmetric encryption systems (also known as *secret-key encryption systems*), the same key is used for encryption and decryption. In asymmetric encryption systems, different keys are used for encryption and decryption. Perhaps the most important class of asymmetric encryption systems are public-key systems, in which the security of one key can be maintained even if the value of the other is made public. Traditional military encryption systems are based on secret-key technology. Public-key encryption systems are often used in commercial applications for authentication and exchange of a randomly generated secret key at the beginning of a transaction, and are beginning to find their way into similar military applications. Secret-key encryption systems tend be less computationally intensive than public-key systems, favoring their use for large exchanges of data.

There are three main forms of encryption: *bulk encryption, message encryption*, and *message-content encryption*. In bulk encryption, the encryption of all data on a link on which transmission is continuous prevents both unauthorized reception and traffic analysis. This form of encryption is used by most trunk communications systems, and is sometimes referred to as *trunk encryption*. As well as protecting against interception, this form of encryption protects against deception.

In message encryption, individual messages on a link are encrypted including both message header and contents. This form of encryption is used in most CNR systems. While the contents of the messages are protected from interception, message encryption does not prevent traffic analysis. If synchronization is achieved through a preamble, no protection is provided against deception by delayed replay of previous traffic.

In message-content encryption, the bodies of messages are encrypted, leaving message headers in plaintext. This form of encryption is often used in packet-switching systems. It has the advantage that intermediate switches and routers are not required to have attached cipher equipment. The disadvantage, however, is that detailed information that can be used in traffic analysis is provided in plaintext. If synchronization is achieved through a preamble, this type of encryption does not protect against deception by delayed replay of previous traffic.

2.6 Channel Coding

Channel coding [5] is used to enhance digital signals so that they are less vulnerable to such channel impairments as noise, fading, and jamming. To do this, some form of *error protection*, or *error detection and correction* is utilized. The numbers of errors at the output of the receiver can be reduced by adding additional information to the message to increase the receiver's decision-making ability. This additional, redundant information increases the overhead that is included in the transmission, but increases the probability that the information will be error-free. There are two main types of channel coding techniques: *block codes* (e.g., parity, cyclic, Hamming, Golay, BCH, and Reed-Solomon codes) and *convolutional codes* (e.g., Viterbi codes).

Once an error has been detected, it must be corrected. There are two main error correction techniques:

- *Automatic repeat requests (ARQ)*. This is the simplest strategy for the correction of errors. Once an error has been detected, the receiver requests the transmitter to retransmit the portion of data that was in error. This can be a very powerful mechanism for the correction of errors, but can produce very low data rates in high-noise environments.

- *Forward error correction (FEC)*. In this method the receiver is sufficiently sophisticated to be able to correct errors itself without reference to the transmitter. It does this by using additional information transmitted along with the data and employing one of the error detection techniques discussed in the last section. The receiver can correct a small number of the errors that have been detected. If the receiver cannot correct all detected errors, the data must be retransmitted.

2.6.1 Block Codes

One of the most common and simplest forms of block error detection code is through the use of a *linear block code*, where additional check bits (*parity bits*) are transmitted with the data at the end of an existing code. The transmitter places these additional bits at the end of blocks of data to help the receiver decide whether each block has been received correctly. On reception, the elements are added and an error is indicated by an output of 0 (for odd parity) or 1 (for even parity). The parity bit is able to detect single errors as well as detect any error patterns that contain an odd number of errors. Parity check bits are widely used in computer communications and information storage.

While the use of a parity bit can detect odd numbers of errors, the receiver does not know which bit is incorrect and therefore cannot rectify the error. More sophisticated codes can be developed to enable the receiver to detect and correct errors, although it can normally detect more than it can correct. In block coding this means using more than one dimension.

A simple two-dimensional block code is the *product code*. In a simple product code the data is arranged in a rectangular array and parity bits used to check both the rows and columns. If there is only one error, the receiver cannot only determine which bit is incorrect but can also correct it. However, the scheme is more expensive in terms of overhead and the information rate is only $12/20$ or 60% of the transmission rate, compared to $3/4$ or 75% for a single parity bit per 3-bit code.

Multiple errors can be detected and corrected by expanding on the two-dimensional product code and using multidimensional vector space, which is divided into smaller vector spaces. However, even more overhead is required and the information rate is even less than the transmission rate.

Cyclic codes are an important subclass of linear block codes, where a new code word in the code can be formed by shifting the elements along one place and taking one off the end and moving it to the beginning. Instead of being generated by a matrix, a cyclic code is generated by a polynomial so that the codes are sometimes called *polynomial codes*. Importantly, cyclic codes have a structure that makes it possible for the encoding and decoding to be performed by simple feedback circuitry.

2.6.2   Convolutional Codes

A *convolutional code* extends the concept of a block code to allow memory from block to block. Each encoded symbol is therefore a linear combination of information symbols in the current block and a selected number of preceding blocks. Therefore, for example, if the final output is a 1 followed by a 0, then these two digits could only have been arrived at by via a certain sequence of 0s and 1s preceding them. The longer the sequence, the easier it becomes for the receiver to detect where the received sequence deviates from a possible sequence and to correct one or more errors. Decoding of convolutional codes is based on the principle of the *Viterbi decoding algorithm* or *sequential decoding.*

2.7   Modulation

The digital and analog signals we have discussed so far are baseband signals, that is, their bandwidths extend from 0 Hz. In order to pass these frequencies over a transmission channel, the baseband frequencies must be translated to a frequency range that is able to pass over the channel. The shifting or translation of a signal from one frequency band to another is accomplished by the process of *modulation,* in which the baseband signal is impressed in some way on a carrier signal at a higher frequency such that one or more of the characteristics of the carrier signal are altered.

2.7.1   Analog Modulation

The three characteristics of a sinusoidal carrier signal that can be altered are the amplitude, the frequency, and the phase. This section describes the three corresponding modulation methods: *amplitude modulation* (AM), *frequency modulation* (FM), and *phase modulation* (PM).

2.7.1.1   AM

As its name implies, AM uses the modulating signal to vary the amplitude of a carrier-wave frequency. In the simplest form of AM the carrier amplitude is switched on and off in accordance with some agreed code sequence such as the Morse code. Figure 2.12 shows the waveform of a sinusoidally modulated wave where a 1-kHz sine wave has amplitude-modulated a 10-kHz carrier.

Figure 2.13 shows the frequency domain representation of a carrier frequency f_c modulated by a single sine wave of frequency f_m. The original frequency f_m is translated by the modulation process to become two side

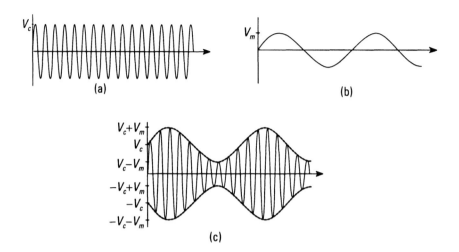

Figure 2.12 The AM waveforms: (a) a 10-kHz carrier modulated by (b) a 1-kHz signal to produce (c) the modulated waveform.

frequencies around the carrier, each of which is half the size of the original f_m, that is, the information contained in f_m has been translated (duplicated) to sit either side of the carrier.

The modulating signal waveform is rarely sinusoidal, but can be viewed as the sum of a number of sinusoids, in which case each component frequency of the modulating signal produces corresponding upper and lower side frequencies in the modulated wave, and the modulation envelope will have the same shape as the modulating waveform. The result is that, instead of a single side frequency above and below the carrier, a band of frequencies (in side frequency pairs) is produced above and below the carrier. The band of side frequencies below the carrier frequency is known as the *lower sideband* and the band above the carrier frequency forms the *upper sideband*. The frequency domain representation of sidebands is shown in Figure 2.14(a).

Double-sideband AM. The power consumed in the carrier transmits no information and is effectively wasted. It would make sense, therefore, to transmit only the sidebands, which do contain the information. When both sidebands of a modulated carrier are transmitted without the carrier, the transmission is called *double-sideband suppressed-carrier* (DSBSC) or, more commonly, DSB. The waveform for a DSB transmission is shown in Figure 2.14(b).

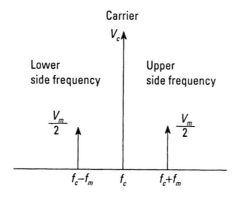

Figure 2.13 Frequency representation of an AM waveform where a carrier with frequency f_c is modulated by a single tone of frequency f_m.

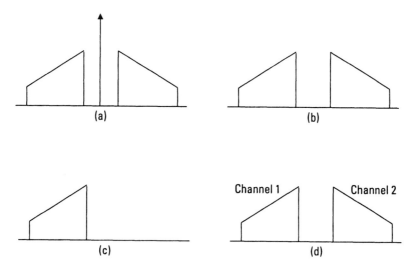

Figure 2.14 Frequency-domain representations of the (a) AM, (b) DSB, (c) SSB (LSB), and (d) ISB waveforms.

Single-sideband AM. Suppressing the carrier in DSB will increase the efficiency of the transmission, since only the sidebands are amplified. Some inefficiency remains, however, since the transmission bandwidth must still be twice that of the message bandwidth due to the transmission of both the upper and lower sidebands. These two sidebands are redundant since all the necessary information can be conveyed by only one sideband. When only one sideband is transmitted, as illustrated in Figure 2.14(c), the transmission is called *single sideband* (SSB), or *upper sideband* (USB), or *lower sideband* (LSB). This reduces the bandwidth of the modulated wave by half and makes it equal to that of the baseband signal. All of the transmitted power then moves toward transmitting the baseband information.

Independent-sideband AM.   Channels can be spaced closer together; hence, further bandwidth economy can be achieved by the use of *independent-sideband* (ISB), where two modulating signals each modulate the same carrier frequency to occupy the upper and lower sidebands, respectively. As illustrated in Figure 2.14(d), the ISB waveform is similar to the DSB waveform, except that each sideband is independent.

2.7.1.2   FM

The distinct disadvantage of all forms of AM is that any noise present in the channel will have an additive effect and will appear as additional modulation on the carrier. This noise is impossible to remove and will be demodulated as part of the received signal. In FM, the amplitude of the waveform remains constant, but its frequency is varied in accordance with the instantaneous value of the modulating signal, as illustrated in Figure 2.15, which shows the effect on the carrier of modulation by the dotted sinusoidal waveform. Any amplitude variations resulting from noise in the channel can be eliminated before demodulation without affecting the information contained in the frequency variations.

An FM receiver has the ability to suppress the weaker of two signals that are at or near the same frequency. This ability to pass the stronger signal is called the *capture effect.* In AM, both signals would be heard at the receiver

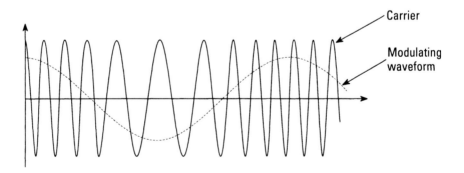

Figure 2.15 A sinusoidally modulated FM wave with the modulating waveform superimposed.

as the modulation on both signals would add and the demodulated effect would be the addition of both audio signals. In FM, only one signal (the stronger) will be heard, which has the advantage of ignoring any interference at or near the same frequency.

FM has a number of advantages over AM. An FM system has a much higher dynamic range (allowable range of modulating signal amplitudes). The FM transmitter is more efficient since the amplitude of an FM wave is constant and each stage in the transmitter can be operated in an optimum way. FM systems are much less susceptible to noise since most noise in the transmission path will be additive and most amplitude variations in the received signal will be ignored by the receiver. An FM receiver also has the ability to suppress the weaker of two signals simultaneously present at its antenna at or near the same frequency.

The main disadvantage of FM is, of course, the much wider bandwidth (perhaps 7 to 15 times wider than AM) required to achieve the signal-to-noise ratio improvement. Since FM requires a wider bandwidth, higher frequencies must be used so that reception is normally limited to line-of-sight. For mobile applications the capture effect may also be a disadvantage when a mobile receiver is near the edge of the service area and it may be captured by an unwanted signal or a noise voltage.

2.7.1.3 PM

PM is very similar to FM except that, instead of the frequency, the instantaneous phase of the carrier is at a rate proportional to the modulating frequency and by an amount equal to the amplitude of the modulating signal. Again, the carrier amplitude remains unaltered. Figure 2.16 illustrates the phase-modulated waveform produced from a carrier and a sinusoidal waveform. PM is rarely used in practice, except as an intermediate step in the production of FM.

2.7.2 Digital Modulation

For the transmission of data there are three versions of these modulation methods that are the simplest forms of AM, FM, and PM: *amplitude-shift keying* (ASK), *frequency-shift keying* (FSK), and *phase-shift keying* (PSK), respectively.

2.7.2.1 ASK

ASK is the simplest form of AM, in which the carrier is turned off and on as illustrated in Figure 2.17. ASK modulation was the first technique developed

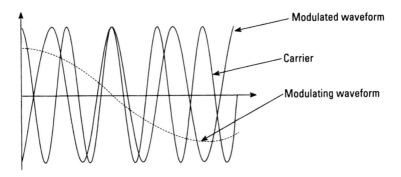

Figure 2.16 PM illustrating the carrier, the modulating waveform, and the resultant modulated waveform.

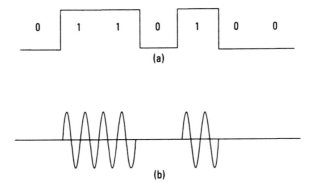

Figure 2.17 ASK illustrating (a) the original digital waveform, and (b) the ASK time-domain waveform.

for the transmission of digital information, but it is not in extensive use today since it suffers from the same significant distortion of the carrier amplitude which is associated with all AM systems.

2.7.3 FSK

The solution to amplitude distortion of ASK is to take advantage of the noise immunity offered by FM systems by using the different voltage levels of the mark and space to displace a continuous carrier by equal amounts above and below the assigned frequency. This technique is known as FSK. An illustration of FSK is given in Figure 2.18.

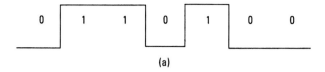

(a)

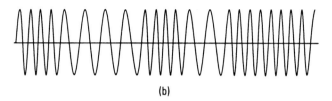

(b)

Figure 2.18 FSK illustrating (a) the original digital waveform, and (b) the FSK time-domain waveform.

2.7.4 PSK

In PSK, digital information is transmitted by shifting the phase of the carrier among several discrete values. For a binary sequence the phase is normally switched between 0° (logic 1) and 180° (logic 0). A variation of PSK is called *differential phase-shift keying* (DPSK) where a phase reversal takes place for each logical "1" whether the symbol before was a logical "0" or a "1." No phase change occurs at the incidence of a logical "0." This helps the receiver and prevents it from confusing a "0" as a "1" and vice versa. Figure 2.19 illustrates PSK and DPSK.

Still, further increases in information rate can be achieved within the smaller bandwidth and lower power requirements of PSK. The rate can be doubled by using *four-phase* or *quadrature PSK* (QPSK). In QPSK the possible phase shifts are 45°, 135°, 225°, and 315° so that 2 bits of information can be indicated for each phase. Figure 2.20 illustrates the polar diagram for a QPSK system compared with a two-phase PSK system.

Hybrid techniques can be formed by combining ASK with PSK to produce polar diagrams that have codes represented not only by varying phases, but also by varying amplitudes. Techniques that combine phase and amplitude changes are called *quadrature amplitude modulation* (QAM).

The precise bandwidth requirements for digital modulation will vary depending on a number of factors, but, for planning purposes, ASK and PSK require the same RF bandwidth, which is equal to the symbol rate. FSK requires RF bandwidth that is 1.5 times the symbol rate.

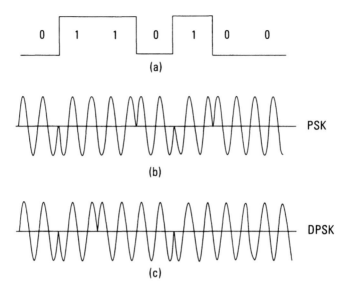

Figure 2.19 PSK illustrating (a) the original digital waveform, (b) the PSK, and (c) DPSK time-domain waveforms.

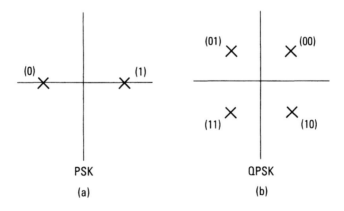

Figure 2.20 Polar diagrams for (a) PSK, and (b) QPSK.

The preferred technique for most digital radios is PSK, due to its bandwidth and power efficiency. However, many in-service radios have been designed for analog operation using FM. In these radios it is natural to use FSK as a digital modulation method; even the data rate is lower for the same bandwidth than that possible using PSK.

2.7.5 Modems

The modulation and demodulation of the digital signal to be transmitted over line or radio is conducted by a *modulator/demodulator* (modem). There are three main types of modem for use in telecommunications.

- *Baseband modem.* A baseband modem digitizes the analog signal and produces *direct-current* (dc) pulses for direct transmission to line. Signaling speed and distance are limited by distortion of the pulses. Baseband signals also cannot be transmitted through systems with amplifiers and other devices that only operate with *alternating-current* (ac) signals. For these systems, the baseband signal must be used to modulate a carrier in a higher frequency range at either VF or RF.

- *VF modem.* A VF modem performs the digital modulation but uses the dc pulses to modulate a VF carrier (in the 1–2 kHz range) using a modulation technique such as ASK, FSK, or PSK. Modulation up to VF frequencies allows transmission over existing telephone lines or radio channels, which are designed to accept voice frequencies with bandwidths of approximately 3.1 kHz.

- *RF modem.* If transmission over other types of medium is required, modulation is performed to other frequency ranges. For example, transmission over radio requires the use of an HF, VHF, or UHF modem.

2.8 Multiplexing

Multiplexing schemes combine the messages from several information sources to be transmitted as a complete group over a single transmission facility with provision at the receiver(s) for separation (*demultiplexing*) back to the individual messages. Two generic forms of multiplexing are of interest: *frequency-division multiplexing* (FDM) and *time-division multiplexing* (TDM).

2.8.1 FDM

FDM is directly applicable to analog sources and forms a composite signal by "stacking" several information channels side-by-side in frequency before modulating a main carrier in some conventional manner. Recovery of the individual messages after reception and initial demodulation is accomplished by bandpass filtering and frequency selection of the channels.

The most common means of obtaining the channels for FDM systems is to use SSB modulation techniques to achieve a minimum bandwidth for each channel. The subcarriers for each channel are chosen so that, after modulation, the channels are stacked side-by-side in the frequency domain. At the receiver, the main carrier is demodulated and each channel of the multiplexed group individually separated out by channel filters located at the subcarrier frequencies.

2.8.2 TDM

TDM is a logical extension of pulse modulation methods and involves interleaving in time the narrow pulses of several ordinary pulse modulation signals, such as PCM or delta modulation, to form one composite digital signal for transmission. Demultiplexing of the time-multiplexed pulse streams at the receiver is accomplished by gating appropriate pulses into individual channel filters. Transmission over radio links would require further analog modulation by either an FSK or PSK modulator.

2.9 Frequency Spreading

In *frequency-spreading*, or *spread-spectrum* techniques the transmitted signal occupies a bandwidth much greater than that of the message signal it conveys. Additionally, the bandwidth is determined by a prescribed *spreading* algorithm, and not by the message waveform itself. There are two main types of spread-spectrum techniques: *direct-sequence spread spectrum* (DSSS) and *frequency hopping* [6].

There are a number of significant advantages to the use of spread-spectrum techniques. Transmissions are concealed since the spectral density of a spread-spectrum transmission may be less than the noise spectral density at the receiver. Despreading at the receiver will spread unwanted signals, thereby "pulling" the wanted signal from the noise. This improvement in signal-to-noise ratio due to the despreading process is called *process gain*. Any spread-spectrum transmission other than the intended one will be seen by the receiver as noise. Spread spectrum techniques are therefore of particular interest in military communications due to their ability to share bandwidth, reject interference and have a low probability of detection.

2.9.1.1 DSSS

A DSSS system spreads the transmitted spectrum as shown in Figure 2.21 by modulating the baseband signal with the digital code sequence produced by the pseudonoise code generator.

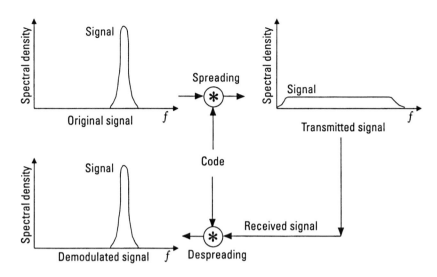

Figure 2.21 DSSS.

When a jammer or other single-channel transmitter introduces the interference spike shown in Figure 2.22, the receiver despreads the signal but, by the same process, spreads the interference. The result is a signal that is harder to detect and unable to be jammed by a spot jammer.

Figure 2.23 shows that a spread-spectrum transmission is also not affected by barrage jammers, or by other spread-spectrum transmissions. Again, any signal not originally spread by the transmitter will not be despread at the receiver, but will be spread to reduce its impact on the wanted signal. So, the wanted signal will be despread and interference will be further spread, increasing the process gain of the receiver. It should be noted that the spreading code is not intended to provide encryption although, if properly managed, a degree of privacy would be provided by the DSSS system.

DSSS modulation techniques have two principal advantages: The spreading of the transmitted signal reduces the power in any one channel below the noise floor of conventional receivers; and unwanted carriers (spread or otherwise) are rejected by the spread-spectrum receiver.

2.9.1.2 Frequency Hopping

As illustrated in Figure 2.24, frequency hopping involves the periodic changing of the frequency associated with a transmission. Successive frequencies are determined by a pseudorandom sequence. A frequency-hopping signal may be regarded as a sequence of modulated pulses with pseudorandom

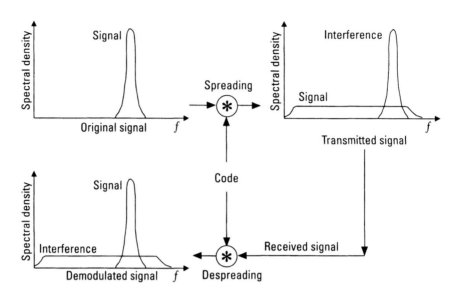

Figure 2.22   DSSS in the presence of single-channel interference.

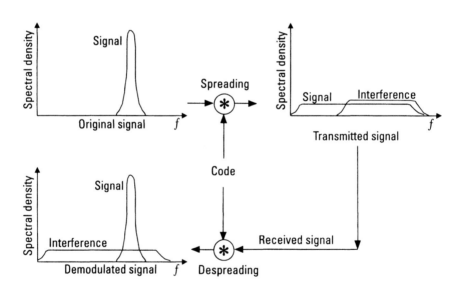

Figure 2.23   DSSS in the presence of broadband interference.

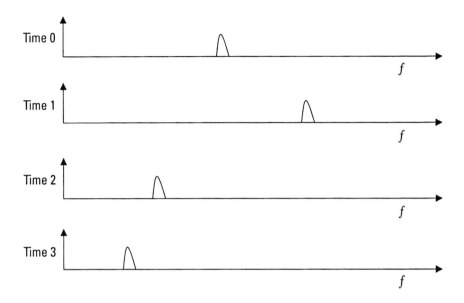

Figure 2.24 Illustration of frequency hopping.

carrier frequencies. Hopping occurs over a frequency band that includes a number of frequency channels. Each channel is defined as a spectral region with a center frequency that is one of the possible carrier frequencies and a bandwidth large enough to include most of the power in a pulse with the corresponding carrier frequency. The bandwidth of a frequency channel is often called the instantaneous bandwidth.

The effectiveness of frequency hopping against a sophisticated jammer depends on the randomness of the hopping pattern but these benefits are potentially neutralized by a frequency-follower jammer. To be effective against a frequency-hopping system, the jamming energy must reach the target receiver before it hops to a new set of frequency channels. Thus, the faster the hopping rate, the more protected the system against a follower jammer.

The main advantage of frequency hopping spread spectrum is in providing interference immunity over bandwidths that are significantly larger than direct-sequence systems (due mainly to the development of frequency synthesis techniques). Consequently, larger processing gains are possible with frequency hopping. Another advantage is that frequency hopping has inherent flexibility in frequency allocation. It is possible to use contiguous and noncontiguous frequency bands utilizing the frequency spectrum in the most efficient way.

2.10   Multiple Access

A channel connecting two points can be shared by multiple users by multiplexing and demultiplexing at each end of the channel. Multiplexing is not an effective way to share a radio channel, however, because the users will not necessarily be in the same location. The techniques used to serve a number of dispersed stations are known as *multiple-access techniques* and are an extension of the principles of multiplexing. There are three most commonly employed multiple-access techniques: *frequency-division multiple access* (FDMA); *time-division multiple access* (TDMA); and *code-division multiple access* (CDMA). The major distinction between these techniques are that, in FDMA, multiple users share a portion of the available channel bandwidth on a continuous basis, whereas, in TDMA, the entire channel bandwidth is used by each of the users, but only for a prescribed time interval. For CDMA, the entire bandwidth is used continually by all users who are separated by the spreading code. These techniques are discussed in more detail in Sections 2.10.1 through 2.10.3.

2.10.1   FDMA

In FDMA, a portion of the electromagnetic spectrum is allocated to each transmitter, which can transmit in its allocated channel all the time. Allocation of frequencies can be fixed or on-demand. When used as a multiple-access technique, frequency hopping (see Section 2.9.1.2) is also a form of FDMA. Each transmitter can transmit all the time by hopping to frequencies selected by the pseudorandom code.

The advantages of FDMA are that no central control station is required unless capacity is demand-allocated, that close to 100% of the available channel capacity can be used, there is no loss of data due to one station overtransmitting another, and there is no delay introduced by the channel. The disadvantages of FDMA are that the allocation of capacity between transmitters is relatively inflexible and that it is relatively difficult for one station to receive data from more than one transmitter.

2.10.2   TDMA

In synchronous TDMA, a fixed-length, periodic time slot is allocated to each transmitter. Timing synchronization requires either a central control station or regular transmissions from all stations. Each transmitter has available the whole channel capacity during its time slot, and must remain silent at all other times. Allocation of time slots can be by fixed assignment or on-demand. The advantages of synchronous TDMA are that it is relatively easy for one station to monitor transmissions in all time slots, that it is possible to use close to 100% of the available channel capacity if there is either a single transmitter or single receiver, that there is no loss of data due to one station over-transmitting another, and that there is a fixed upper bound on delay. The disadvantages are that timing synchronization between stations is required, that guard intervals reduce channel capacity when multiple transmitters and multiple receivers are used, and that allocation of capacity between transmitters is relatively inflexible.

Carrier-sense multiple access (CSMA) techniques are a form of asynchronous TDMA in which there are no fixed time slots. A station wishing to transmit checks first that no other station is currently transmitting. If the channel is free, the station transmits; if not, it waits a random period of time and tries again. If two stations inadvertently transmit simultaneously, both recognize the collision, cease transmission, and wait a random period of time before trying again to access the channel. The advantages of CSMA are that it is relatively easy for one station to monitor all transmissions on the channel, that no central control station is required, and that the allocation of channel capacity is very flexible. For the transmission of data, the disadvantages of CSMA are that the best throughput that can be achieved is approximately 50% of the available channel capacity, that data is lost due to one station overtransmitting another, and that there is no fixed upper bound on delay.

2.10.3 CDMA

CDMA allows a wideband channel to be shared by a number of narrowband sources by spreading their transmissions over the whole band using DSSS. By using a different spreading sequence for each transmitter, multiple access is achieved. The advantages of CDMA are that no central control station is required unless capacity is demand-allocated, close to 100% of the available channel capacity can be used, there is no loss of data due to one station over-transmitting another, and there is no delay introduced by the channel. The disadvantages of CDMA are that the near-far effect makes it infeasible to have more than one transmitter and more than one receiver operating simultaneously, the allocation of capacity between transmitters is relatively inflexible, and it is relatively difficult for one station to receive data from more than one transmitter.

2.10.4 Time Hopping

Data can also be encoded in variations in the length of intervals between the transmission of very short impulses. These impulses may be as short as 1 ns, leading to a total transmission bandwidth of 1 GHz. The advantages of time hopping include the potential to remove any need for frequency management. The disadvantages include a high degree of interference between time-hopping signals and conventional communications systems that share the same bandwidth.

2.11 Transmitters

At the transmitter, the information signal is used to modulate a suitable carrier frequency, which translates the signal to an allocated part of the frequency spectrum [7]. The result is then amplified to the power level required for transmission. In an AM transmitter (Figure 2.25) the carrier is generated by a very stable local oscillator and then amplitude-modulated before being amplified and applied to the antenna. An FM transmitter has a very similar block diagram except that a preemphasis network is included to artificially boost the high-frequency components to protect them against the nonlinear noise that is introduced in the FM demodulation process.

2.12 Receivers

At the end of the communications channel is the receiver whose function is to receive the signal and present it to the sink in an appropriate form. To do this the receiver must be able to select the desired signal from all the signals present at the antenna, amplify and demodulate, and present it to the source

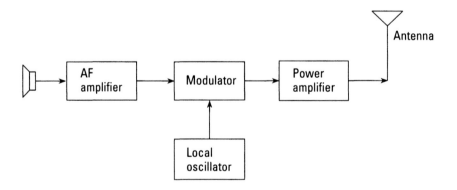

Figure 2.25 Block diagram of a low-level AM transmitter.

in the appropriate format. The functions of a receiver are fairly generic, with most receivers having the same basic form. The precise design of any particular receiver will depend on what is required of it, such as the amount of amplification and the type of demodulation [8].

The basis of the modern superheterodyne receiver is the conversion of the wanted signal frequency into a constant frequency known as the *intermediate frequency* (IF). It is at this IF that most of the gain and the selectivity of the receiver is provided without affecting the bandwidth of the receiver. The basic block diagram of the superheterodyne radio receiver is shown in Figure 2.26. The ganged-tuning arrangement ensures that the RF stage and the local oscillator are tuned together so that they are always separated by the same amount (the IF). When they are combined together in the mixer, the wanted signal is converted in frequency to the IF.

The superheterodyne receiver of Figure 2.26 is suitable for AM signals. The FM receiver has a very similar system diagram except, of course that, due to different demodulation requirements, the detector circuit will be different. Additionally, an FM receiver has two further circuits: a *limiter* to remove amplitude variations from the carrier before demodulation, and a *deemphasis network* that redresses the preemphasis network's nonlinear amplification of the high audio frequencies.

2.13 Transmission Lines

Transmission lines are used widely to connect pieces of equipment and particularly to connect the transmitter and receivers to their respective antennas. Ideally, transmission lines should be able to cope with wide bandwidths (high capacities), have low attenuation so that the signal reaches the end of the transmission line at the same strength as it entered, and be matched to its

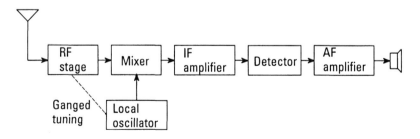

Figure 2.26 Block diagram for a superheterodyne receiver.

termination to allow for maximum power transfer. They should also not be affected by crosstalk or noise and interference, so that signals are not affected by signals in another transmission line, or from other sources of noise and interference; and have low radiation so that signals on the line cannot be intercepted at any reasonable distance. Additionally, transmission lines should have low dispersion so that all frequencies have the same velocity of propagation and arrive together at the end of the transmission line, be able to cope with the required power levels, be easily handled and installed, and maintain personnel safety.

There are two main types of transmission line: *balanced* and *unbalanced*. At any one time, balanced lines have voltages (currents) on the two conductors that are equal and opposite relative to Earth. Unbalanced lines have a positive or negative voltage on one conductor and the other conductor is earthed (at 0V).

2.13.1 Balanced Transmission Lines

2.13.1.1 Two-Wire Line

Two-wire lines consist of two parallel conductors maintained at a fixed distance apart. There are two main types: *open two-wire line* and *insulated two-wire line*.

Open two-wire line. In an open two-wire line, the two parallel conductors provide the balanced transmission line. The balance is easily disturbed, however, by nearby metallic objects, and large radiation losses can be obtained. The balance is maintained by keeping the two wires apart by means of insulating bars, or spacers.

Insulated two-wire line. Open two-wire line is difficult to install due to the requirement to maintain the spacing between conductors (and therefore maintain the balance). Insulated two-wire line uses a solid dielectric between the conductors instead of air, which provides a cable that can be installed more easily. This type of flat ribbon cable finds common application in connecting home television antennas to the set.

2.13.1.2 Twisted Pair

Twisted pair consists of two insulated copper wires twisted together to maintain the wires at a fixed distance apart. The twisting also limits radiation from the wires by aiding in the cancellation of the electric and magnetic fields and balances them against the effects of any induced radiation. Further reduction in both effects is often obtained by placing a screen around the twisted pair [often called shielded twisted pair (STP)]. Twisted pair has relatively high losses, high radiation, and low bandwidth, but it has the advantages of being very cheap and easy to install. Recent advances in technology have seen a resurgence in the use of twisted and shielded twisted pair as a transmission medium for local area networks in which significant data rates (100 Mbps) can be achieved over relatively short distances.

2.13.1.3 Multicore Cable

Multicore cable is made up of up to approximately 50 twisted pairs that are shielded by a foil sheath. Each twisted pair is color-coded and the cable is sometimes strengthened by the inclusion of nonconducting elements. Multicore cables operate at low frequencies (normally megahertz), can be used to carry moderate voltages, and can be used over limited distances (normally kilometers). Because of the lack of a dielectric between the pairs, multicore cable is very susceptible to crosstalk between pairs, particularly as the cable ages and the insulation around each pair begins to deteriorate.

2.13.2 Unbalanced Transmission Lines

Coaxial Cable

Coaxial cables consist of two concentric conductors separated by a dielectric, normally teflon or polyethylene. The majority of electromagnetic field is restricted to the cable and most radiation losses are eliminated. The outer conductor also shields the inner wire from any induced radiation from any RF sources in the vicinity. The dielectric operates as a capacitor modifying the characteristic impedance. Coaxial cables can be used for frequencies up to 3 GHz and have common characteristic impedances of 50Ω, 75Ω, 300Ω, or 600Ω.

2.13.3 Advantages and Disadvantages of Copper Transmission Lines

Balanced and unbalanced copper-based transmission media have been used for as long as communications systems have been in existence. They are based on proven technologies and can be installed at low cost. They have the additional advantages of being able to carry high powers and are easy to use, join, and terminate.

However, copper-based transmission media have a number of significant disadvantages. They are heavy, particularly when deployed in large numbers in cable ducts; they suffer from significant crosstalk and large losses when used over more than short distances. They also suffer from interference from RF sources along the path. Transmission quality is reduced due to the high-frequency limitations of copper cables. As bandwidth requirements increase, so must the operating frequencies at which component specifications are increasingly difficult, and therefore more expensive, to achieve.

These disadvantages become very important for medium- to long-distance communications requiring high bandwidth, and copper cable is now rarely used due to the high cost of providing a large number of cables and the many repeaters required. The preferred transmission medium is based on optical-fiber technology.

2.14 Optical Fibers

In transmission lines, the signals travel as currents and voltages on two conductors. *Waveguides*, as their name suggests, provide a medium for signals to travel as electromagnetic waves inside a metal tube, which is normally rectangular. Waveguides of this form are usually only used at SHF and above. We will not consider them here, but we will briefly outline a much more common waveguide—optical fibers [9].

An optical fiber is constructed by enclosing a thin glass fiber core in a glass cladding and surrounding the result in a protective jacket. Electrical signals are translated into light pulses by modulating a laser, and are detected at the receiver by photoelectric diodes. A waveguide for the optical frequencies is provided by the difference between the refractive indexes of the cladding and the glass core. The size of the core and cladding can differ, but commonly the cladding is 125 μm in diameter and the core is between 8 and 62.5 μm.

2.14.1 Advantages of Optical Fibers

Optical fibers provide large bandwidth and can carry many times more signals than copper cables. Because there is such a low loss associated with the propagation of light in the fiber, optical fibers also provide extremely reliable long-distance communications. The fibers are light and small in size, making them ideal for installing in buildings and existing cable ducts. Since the information is contained in light energy, there is no RF radiation from optical fibers, nor are they affected by noise of interference in the areas through which they are run. The lack of radiation provides very secure communications, while the immunity to noise and interference makes them ideal for use in heavy machinery workshops and similar locations. Since the signal in the fiber is not electrical, there are no earthing problems associated with installation.

2.14.2 Disadvantages of Optical Fibers

Although the raw material for optical fibers is plentiful and therefore cheap, the propagation channel for the light is critical and is very hard to repair or join. The repair process requires that the core be precisely aligned and must ensure that the two cables abut perfectly and no dust or grime is allowed to intrude into the joint. Multiplexing is difficult with optical fibers so they are mainly used for high-capacity, point-to-point links. Similarly, the cables are very hard to terminate or to split to allow switching or tapping, compared with wire cables, which can be joined with a simple twist. If repair of a wire cable that had a similar capacity as an optical fiber was to be considered, however, the repair of the fiber has a similar, if not better, repair time. The cost of fiber-optic cable can be greater for short distances, but for distances greater than several hundred meters, fiber-optic cables are cheaper for similar capacity since the copper cable requires many more repeaters.

2.15 Antennas

An antenna (or aerial) [10] is a device for radiating or receiving electromagnetic waves. It is used to connect the transmitter and receiver to the communications channel, through which electromagnetic waves will propagate. There is little fundamental difference between transmitting and receiving antennas and the same antenna is commonly used for the both purposes. It is relatively simple to construct an antenna since any length of conducting wire that is carrying a current will radiate an electromagnetic wave.

As illustrated in Figure 2.27(a), a dipole antenna is constructed by opening the legs of a balanced transmission line. Figure 2.27(b) shows a monopole antenna, which can be constructed by using only the upper arm of the dipole and placing it above a reflecting ground plane so that its image appears as the other side of the dipole. A monopole antenna is fed by an unbalanced transmission line and has obvious attractions for mobile operation since it is half the size of the dipole.

2.15.1 Antenna Size

Although any length of wire will radiate, to be efficient, the wire length must be half a wavelength (or some multiple) long so that all of the energy that is

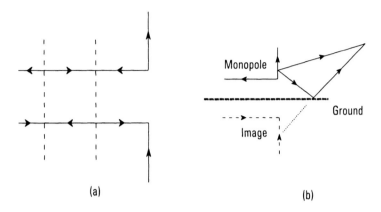

Figure 2.27 Antenna construction for (a) the dipole, and (b) the monopole.

contained in the transmission line is efficiently transferred to the radiated electromagnetic wave. This requirement for efficiency means that the length of an antenna is related to the frequency of operation; the lower the frequency, the larger the antenna. For example, a 2-MHz transmission requires an antenna that is one-half a wavelength, or 75-m long. Yet a 200-MHz transmission requires an antenna that is only 0.75-m long. Higher frequencies therefore allow the use of more mobile antennas.

It follows therefore that an antenna must be erected to be the appropriate frequency in use. Further, if the frequency is changed, the antenna must be taken down, modified in length, and then reerected. In some mobile applications changing the antenna length is not feasible and an antenna tuning unit (ATU) is used to change the electrical properties of the antenna so that it always appears to the transmitter to be an antenna of the correct length. While an ATU protects the transmitter from any power that may be reflected from a short antenna, it must be realized that a short antenna will only ever radiate commensurately less power than one of the correct length. So, while the ATU has some utility, it can never compensate for having an antenna of the correct length.

2.15.2 Radiation Pattern

The radiation pattern of an antenna is an important consideration for its application. Figure 2.28 shows the radiation pattern for a dipole. The maximum radiation is broadside on to the dipole with nothing coming out the ends. To get the three-dimensional picture, imagine the "figure of eight"

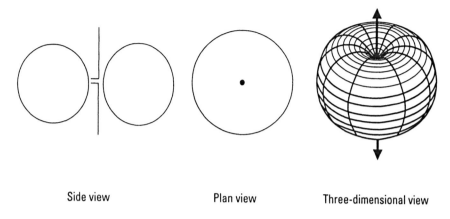

| Side view | Plan view | Three-dimensional view |

Figure 2.28 Radiation pattern for a vertical dipole in free space.

pattern rotating through 360°. The volume swept out by the figure would be toroidal, which, if viewed end-on would appear circular. So, in that plane, the dipole is omnidirectional.

The pattern in Figure 2.28 is for a dipole in free space. Radiation patterns of antennas are significantly affected by the proximity of the ground or other objects. For example, Figure 2.29 shows the radiation pattern for a horizontal half-wave dipole at various heights above the ground. Generally, as the antenna is raised above the ground, more power is radiated upward and communications can be achieved over greater distances.

2.15.3 Gain

An antenna has *gain* if it radiates more power (usually in one particular direction) than another antenna. Gain is normally obtained by making an antenna directional, that is, by altering its radiation pattern so that more of the radiation is transmitted in a certain direction. Antenna gain plays an important part in the ultimate performance of a system and is closely associated with the transmitted power and the receiver sensitivity employed. The maximum power gain obtainable in practice is usually restricted by various factors such as antenna size, weight, and the need for mobility.

One common method of building an antenna with gain is to use some form of antenna array. For example, consider the half-wave dipole in Figure 2.30(a) that has two lobes, A and B. If we redirect lobe A and add it to B, we should double the power radiated in that direction. We can do this by adding

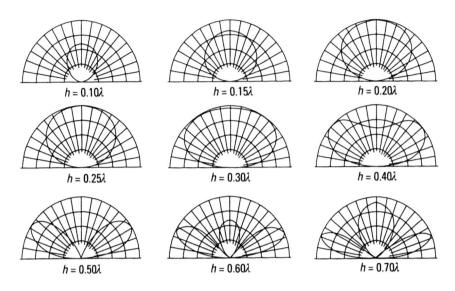

$h = 0.10\lambda$	$h = 0.15\lambda$	$h = 0.20\lambda$
$h = 0.25\lambda$	$h = 0.30\lambda$	$h = 0.40\lambda$
$h = 0.50\lambda$	$h = 0.60\lambda$	$h = 0.70\lambda$

Figure 2.29 Radiation patterns for half-wave dipoles at varying heights above the ground.

a reflector at a suitable distance as illustrated in Figure 2.30(b) or a parasitic director as shown in Figure 2.30(c). Doubling the power gives the antenna twice the gain (relative to a half-wave dipole).

2.15.4 Bandwidth

The bandwidth of an antenna is the frequency range over which the antenna will perform effectively. In the case of dipoles, monopoles, and related antennas, the bandwidth is small. These are known as "resonant" antennas because they respond only to one frequency, which is the frequency at which they are $\lambda/2$ long. So if a dipole's frequency is changed, it must be pulled it down and recut to be $\lambda/2$ at the new frequency if efficiency is to be maintained. If the dipole length is not changed, or when using a mast or whip, the ATU must be retuned as the operating frequency is changed.

Some antennas are *broadband*, and are designed to be efficient over a wide range of frequencies. Examples include the *discone* and the *log-periodic* type of antenna. *Traveling* wave antennas are another class of antenna, which are not resonant and are efficient across a range of frequencies. Their radiation patterns change substantially with frequency, however, so they are not truly broadband.

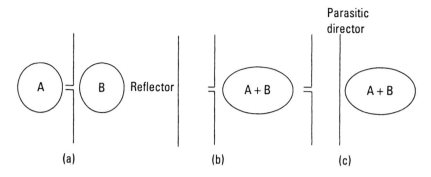

Figure 2.30 Doubling the gain of a (a) half-wave dipole with (b) a reflector, and (c) a parasitic director.

2.16 Propagation

Radio communications techniques are normally categorized in accordance with the method of propagation used. These methods fall into three groups: those which occur in close proximity to the Earth and are called *ground waves*; those which rely on reflections from the atmosphere, called *sky waves*; and those which rely on scattered energy from atmospheric turbulence, called *scattered waves*. A communications system will utilize one of the above mechanisms of propagation [11].

2.16.1 Ground Waves

Figure 2.31 shows the RF energy transmitted by an isotropic radiator that will be received by an antenna as ground waves [12]. Some of the energy will travel directly to the receiving antenna in what are called *direct waves*. Some of the energy will be directed toward the ground, be reflected, and then be received by the antenna. These are called *ground-reflected waves*. The combination of the direct and the ground-reflected waves is called *space waves*. The remainder of the energy will be diffracted around obstacles and ground features and is propagated by *surface waves*.

2.16.1.1   Space Waves

Space-wave communications use high frequencies that tend to travel in line-of-sight. The range of these communications is therefore limited by terrain as well as power. There needs to be sufficient transmitter power to ensure sufficient strength signal at the receiver, but most ground-based radio systems are terrain-limited rather than power-limited. That is, the curvature

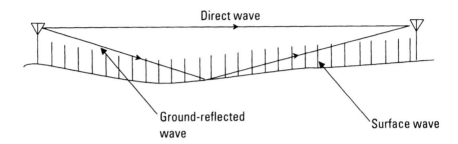

Figure 2.31 Ground-wave propagation.

of the Earth constrains the line-of-sight that is possible between two low antennas. The *radio horizon* can be stated geometrically as:

$$d_{radio} = 4.12\left(\sqrt{h_T} + \sqrt{h_R}\right)$$

where d_{radio} is in kilometers and h_T and h_R are in meters. For example, for two hand-held radios held waist-high (around 1m), the radio horizon is approximately 8.24 km, even though the radios may have sufficient power to operate over tens of kilometers in free space. If a 9-m mast is used for the antennas for each radio, the radio horizon is extended to 24.72 km.

2.16.1.2 Surface Waves

Surface waves, as the name implies, travel along the surface of the ground, with the ionosphere and the ground acting as two sides of a waveguide. The wave is supported by currents flowing in the ground and follows the surface of the terrain. Surface wave is attenuated much more than any other form of wave and consequently has a short range. The attenuation is based on the conductivity of the ground and the polarization of the wave, both of which therefore have an effect on the range of surface-wave propagation.

Surface-wave frequencies are very low, extending only up to approximately 5 MHz. These low frequencies require large antennas and are therefore limited in mobile applications. Additionally, the low frequencies have very low bandwidths.

2.16.2 Sky Waves and Scattered Waves

Communications via space waves are limited to line-of-sight distances. The limit for surface waves is dictated by a number of considerations but is generally around 60 to 70 km. Propagation over longer distances is due to the atmosphere, which is divided into three zones as illustrated in Figure 2.32 and described as follows:

- The *troposphere* extends from the ground to a height of about 11 to 16 km. In the troposphere the air is in permanent motion and meteorological phenomena take place. The influence of the weather prevails, the temperature decreases with height, and cloud formation and convection predominate. There is no ionization of the air.

- The *stratosphere* is located above the troposphere and extends to a height of 40 to 50 km. In the stratosphere the air is still, the temperature is almost constant, humidity is almost absent, and there are almost no perturbations. The stratosphere is not used for communications. The stratosphere ends with the stratopause where there is not enough oxygen molecules to form ozone. Above the stratopause is the mesosphere, which extends from 50 to 80 km, in which the air cools again.

- The *ionosphere* is located about 50 to 60 km from the surface of the Earth where the ionization of the air is appreciable. The lower layer, the D layer, includes the mesosphere. The reflection properties of this

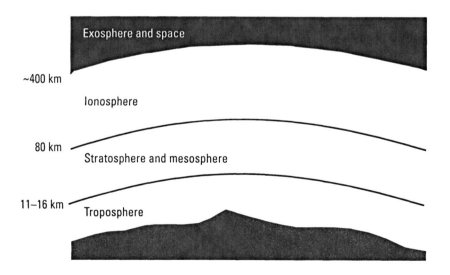

Figure 2.32 Layers of the atmosphere.

region at certain high frequencies are used for very long-distance communications.

Figure 2.33 shows that low-powered RF signals may be received a long way beyond the horizon through the reflection of sky waves from the ionosphere. Energy may also be received as scattered waves from turbulence in the troposphere, or from turbulence in the ionosphere or the ionized tails of meteors.

2.16.2.1 Sky Waves

Sky-wave communication [13] utilizes the ionosphere to refract waves back to Earth, providing a mechanism by propagating over very large distances. Since the height of formation of the ionosphere depends on the intensity of ultraviolet radiation the layers will vary in height from day to night and from summer to winter. Under normal solar radiation conditions the gases present in the atmosphere allow up to four layers of ionization to exist during the day (D, E, F1, and F2). At night, when the solar radiation is reducing, recombination of ions and free electrons predominates and when the solar radiation is at a minimum, the D and E layers effectively disappear and the two F layers combine to form one layer.

Radio waves travel in straight lines in free space, providing that the medium through which they pass has a constant refractive index. Thus, as a wavefront enters the D layer, its path is immediately altered and a slight bending of the wavefront occurs. As the D layer has only a small number of

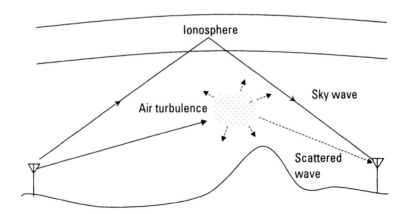

Figure 2.33 Sky-wave and scattered-wave techniques.

free electrons, only LF waves are readily affected and as a consequence, most of the signal energy will be absorbed.

HF waves, however, are not so readily affected and will continue along their original path up to the E layer. As the HF wavefront penetrates the E layer, which has a greater number of free electrons, it begins to follow a gradual curved path. The influence of the free electrons is such that the velocity of the wavefront is slightly reduced, thus causing a refraction of the wave. If there is a sufficient number of electrons, the path of the wavefront is bent back toward the Earth.

If the frequency of the radio waves being transmitted is gradually raised, a frequency will be found beyond which the waves will not be refracted sufficiently to curve their path back to Earth. Consequently, these waves will continue up to the next layer, or, in the case of the F2 layer, into space.

Sky-wave propagation depends, therefore, very much on whether the ionosphere will refract a wave of a particular frequency. In general, the higher the frequency, the easier it is for the wave to pass through the layers without being refracted. At night when there is only the F layer and a weak E layer, the transmitting frequencies must be lower to prevent the wave from passing through both layers and out into space.

The range of sky-wave communications varies and is determined by three main factors: the frequency used, the angle of transmission, and the power of the transmitter. If the frequency is raised, the wave will be refracted by a higher layer and lead to a longer range. The distance between the transmitting antenna and where the first useable sky wave returns to Earth is called the *skip distance*. Similarly, if the propagation angle is lowered, a greater communications range is achieved. The propagation angle is also known as the *take-off angle* and sometimes as the *angle of departure*. A single-hop HF sky wave is limited to ranges of approximately 3,500 km. If the power of the transmission is increased, the range of the transmission is increased due to the occurrence of multiple hops of the transmission. As the wave is refracted back to Earth, it has sufficient power to be reflected back up to the ionosphere where it is again refracted. Worldwide communications are achieved in this manner and ranges of up to 6,000 km are obtained.

The ionosphere suffers a number of variations that will have a significant effect on sky-wave communications. Ionization stops at night, while recombination continues. Daily variations are generally predictable and are accounted for in prediction charts. There is also obviously more ultraviolet radiation at the equator than at the poles, and there is an east-west variation. Additionally, winter and summer frequencies will vary due to variations in the height of the layers. This is due not only to different quantities of ultra-violet radiation but also to the expansion and contraction of the atmosphere. Predictions also need to be for the appropriate month. Additionally, the Sun's intensity varies with an 11-year cycle and sudden variations can occur in solar activity in the form of solar flares and magnetic storms.

Sky-wave communications have the following advantages:

- Over-the-horizon communications are possible.
- Medium distance communications (1,000 km) are available using man-portable and mobile equipment.
- Both net communications and point-to-point links can be established.
- Relatively low-cost antennas and terminal equipment are required.
- The transmission medium (the ionosphere) is difficult to interrupt.
- Sky-wave communications can be degraded in that the same communications system can be used for voice and data at reasonable speeds down to low-speed Morse.

Sky-wave communications have the following disadvantages:

- There is limited available spectrum, leading to a limited number of available channels.
- Channels are available worldwide with a large number of users possible.
- Limited bandwidth is available on each channel, leading to low data rates.
- The transmission medium (the ionosphere) is difficult to predict accurately and contains noise and long-range interference; channel use therefore varies with time.
- Since use of the lower frequencies is desirable, the long wavelengths at sky-wave communications mean that large antennas are required.
- Management is required for efficient use of the ionosphere.

2.16.2.2 Scattered Waves

Scattered-wave techniques are able to make use of the turbulence in the tro-posphere or the ionosphere or ionized meteor tails. Ionospheric scatter and meteor burst techniques are not often used [14]. Troposcatter techniques are, however, used in both commercial and military communications networks [15].

A troposcatter path can vary in length from about 100 to 150 km to almost 1,000 km. Most of the current systems, however, use a path length of between 150 km and 400 km. The maximum path attenuation including deep fades is very high, on the order of 190 to 240 dB. Consequently high transmitted powers, low-noise, sensitive receivers, and high-gain parabolic antennas are normally required for troposcatter communications.

Although a variety of RF power outputs have been used for troposcatter links, normal RF power outputs are 100W, 1 kW, and 10 kW, with the current trend favoring 1 kW to provide path lengths of 10 to 200 km. Owing to the variation in path loss throughout the year, the RF power can sometimes be decreased with a consequent reduction in power consumption and interference. The antennas are usually of the parabolic type with diameters ranging from a few meters up to 30 to 40m. Transportable systems have antenna sizes of 2 to 10m. Receivers with low-noise characteristics are normally employed.

The frequency band of use in practical troposcatter systems is considered to be between about 300 MHz and 5 GHz. Early troposcatter systems provided analog FM communications in the 345- to 988-MHz band. These frequencies cannot handle the bandwidths required for digital communications, so later digital systems operate in either the 1.7- to 2.1-GHz or the 4.4- to 5.0-GHz bands.

Troposcatter systems must accommodate the characteristics of the propagation, which produces frequent deep fades in the received signal. Receiving terminals use a diversity technique in which a number of independent samples of the RF signal are combined in a diversity combiner to obtain a stable output signal. Normally two or four receivers are used to provide diversity techniques incorporating frequency, space, and time diversity. Frequency diversity provides a transmission on several frequencies simultaneously, and the receive terminal selects the strongest signal at its multiple receivers. Space diversity utilizes a number of separate antennas. Time diversity sends the same information over the same link at different times.

Both analog and digital techniques are used in troposcatter links. Digital systems have the advantages of better transmission quality; better protection against interference and jamming; the ability to use error detection and correction; and the ease of interface with other digital systems. However, these advantages may be outweighed by other features (such as wider bandwidths), and all aspects of the problem must be considered before a system is chosen.

Analog troposcatter systems have a traffic capacity of 6 to 120 telephone (4-kHz FDM) channels. Higher capacities are generally not recommended because the propagation mechanism generates nonlinear noise in the channels, thus limiting the bandwidth that can be transmitted. Systems carrying 240 channels have also been implemented at the cost of accepting a lower quality. However, for television transmission the performance would normally be rather poor. Frequency modulation is always used.

Digital troposcatter systems can transmit a bit stream of up to 8 to 12 Mbps. These systems are more sensitive to propagation distortions and their bandwidth becomes too large above 2 to 3 Mbps, so that even at an acceptable path loss, transmission is impossible unless special complex modem techniques are used. Modulation can be by either FSK or PSK.

2.17 Switching

2.17.1 Circuit Switching

A *circuit* is a two-way path for carrying information, either digital or analog. As illustrated in Figure 2.34, efficient use of resources is available using *circuit switching* where each of the users is connected to a central *switch* that provides the capability of interconnecting any two users. The path is in existence only for the duration of the communication and is disconnected at the end.

Figure 2.35 shows how the circuit-switched network is built up. Users are connected to a switch, and switches are interconnected using *trunk* links. The terminology was developed in the commercial network, based on a tree (since the structure was very similar). Major links between switches were trunk links and links from switches to users were called *branch lines*. Additional trunk lines then interconnect each suburban switch with a major switch in the city and these major switches are again interconnected by high capacity trunk links between cities. The resultant network is referred to as the *public switched telephone network* (PSTN).

The advantage of circuit switching is that, once the circuit has been set up, there are no delays and communications between the two ends can continue until the circuit is disconnected. Circuit switching is the most efficient switching technique for voice communications since, to obtain real-time conversations, it is necessary to provide a circuit for the duration of the conversation. Circuit switching is also a very efficient way of transferring large volumes of data between data terminals and has the added advantage of requiring relatively simple protocols.

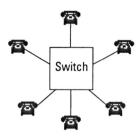

Figure 2.34 A simple switched network with one switch.

Despite this efficiency for voice communications, circuit switching has significant disadvantages for data transfer, since it makes poor utilization of links and equipment and is generally expensive to implement. Also, the set up and clear down times are long in comparison with the time it takes to transmit the data; and the user must continually retry to connect if the other subscriber is absent or engaged.

2.17.2 Message Switching

A *message* is a discrete data communication. Message switching is a store-and-forward concept in which a message with an appropriate destination address is sent into the network and is stored at each intermediate switching point (network node) where its integrity is checked before it is sent onto the next stage of its journey. Switches will duplicate the message if it is for multiple addresses. If the required link is congested or temporarily unavailable, the

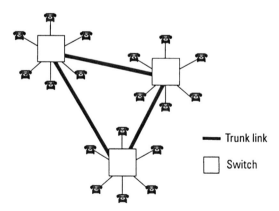

Figure 2.35 Interconnection of circuit switches using trunk links.

message may be stored by the node or sent by an alternate route. Due to this storing and forwarding in message switching networks, the process is often called *store-and-forward* and the network nodes are often called *store-and-forward switches.*

Message switching has an advantage over circuit switching in that it makes good use of the available links and equipment. It does, however, have long response times since messages are stored awaiting available links. Although this is acceptable for traffic for human consumption, the types of delay experienced make message switching undesirable for data communications.

2.17.3 Packet Switching

Packet switching [16] provides both the response of circuit switching and the efficient link utilization of message switching by imposing a maximum length on the transmitted units by splitting all messages into packets up to a few hundred bytes long. Long messages are therefore prevented from blocking links and a system of TDM is forced onto the network. Packet switching is the most economical and cost-effective technique for using data communications links.

In principle, it is possible to pass digitized voice over a packet-switched network. In practice, however, high link speeds are required with large processing capability in the data terminals if an acceptable quality of service (QoS) is to be provided. Packet switches are therefore not usually used for the transmission of voice.

2.17.4 Cell Switching

Cell switching can perhaps be described best in terms of its major instantiation—*asynchronous transfer mode* (ATM) [17]. ATM is a cell-switching and multiplexing technology designed to combine the benefits of circuit switching (constant transmission delay and guaranteed capacity) with those of packet switching (flexibility and efficiency for intermittent traffic) providing support for applications that exchange information in many different formats (e.g., data, voice, image, and video).

ATM's use of small, fixed-length cells offers several important advantages. Because the cell headers are uniform, cell switching can be executed in hardware, allowing greater switching speeds. Delay variability is also more tightly bounded because fixed-length cells have a predictable service rate, and bandwidth reservation is easier. Network scalability is perhaps ATM's most desirable feature.

2.18 Networking and Internetworking

Networks that interconnect buildings (i.e., interconnect LANs) are called metropolitan area networks (MANs) or campus networks. Normally, a MAN is geographically confined to a group of buildings with a common purpose such as a university campus, a factory, or a defense base. MANs (or LANs) can be interconnected by wide area networks (WANs), which normally extend across a city or between cities. Generally, LANs, MANs, and WANs are based on different network technologies. However, the distinction between WAN, MAN, and LAN is becoming blurred as the various technologies converge. In modern networks it is more usual to distinguish between networks (LANs and MANs) and internetworks (WANs).

This section examines networking techniques and looks at methods for extending networks as far as possible. Internetworking techniques are then examined as ways of providing wide-area, scalable networks. First, however, the ISO OSI 7-layer model is described.

2.18.1 ISO OSI 7-Layer Reference Model

Most digital communications systems use procedures arranged in accordance with the International Standards Organization (ISO) Open Systems Interconnection (OSI) reference model. The ISO OSI reference model arranges the procedures for communications for data networks into seven layers or levels. The procedures used at each layer are called protocols. The ISO OSI reference model is illustrated in Figure 2.36.

Throughout the ISO OSI reference model each layer is transparent to the layer above it. This is achieved by each layer receiving a block of information from the layer above it and then adding the information needed for its protocol at one or both ends. The information block from the higher layer is left intact. In the other direction of communication, each layer receives a block of information from the layer below it, strips off the information it requires for its protocol, and then sends the resultant information back up to the next higher layer.

Layers 1 to 3 contain the network protocols. Layers 4 and above contain the user-to-user protocols in which the network takes no part. The network protocols are obviously the most important from a communications point of view. The seven layers are summarized next.

- *Layer 1: Physical layer.* This layer allows the connection, maintenance, and disconnection of physical circuits between different types

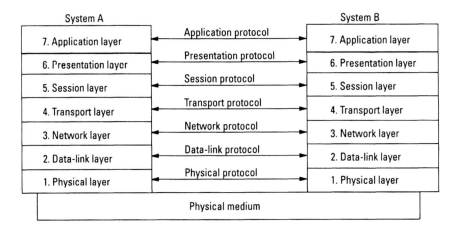

Figure 2.36 Protocol layers for the ISO OSI reference model.

of devices. It provides the mechanical, electrical, functional, and procedural characteristics that enable the data stream to be sent from one terminal to another. This is basic electrical communications that covers the standards associated with plugs, sockets, leads, and so on. Examples are RS-232, RS-449, V.21, V.24, and X.21.

- *Layer 2: Data link layer.* This layer allows point-to-point or node-to-node control. It is often called the frame level because it specifies the formats in which transmissions are embedded in frames. It incorporates error detection and correction and data-flow control providing an error-free link level communications facility to higher levels. *Cyclic redundancy checking* (CRC) is a common method utilized by this layer to detect errors.

- *Layer 3: Network layer.* This layer controls the exchange of data between locations in the network, so that the network is transparent to the data. Routing, flow control, and switching can be provided at this layer. It is sometimes referred to as the packet layer as it defines the packet format and control procedures for the exchange of packets containing both control messages and user data. It multiplexes packets on links and then across the network. X.25 and Internet Protocol (IP) are examples of this type of protocol.

- *Layer 4: Transport layer.* The transport layer establishes and terminates connections between communicating devices, and provides error handling and flow control. Message headers and control messages are added to packets by the transport layer to ensure the integrity of the communications link.

- *Layer 5: Session layer.* This layer determines the rules for establishing and ending a communication and reestablishing a connection if it is interrupted. It determines the right of a device to interrupt another, checks for user authenticity, and keeps track of billing details.

- *Layer 6: Presentation layer.* The presentation layer provides code conversion, compression, and standard layouts for devices and peripherals. This layer's main job is to ensure that the data is "presented" to the application layer in a way that layer can understand.

- *Layer 7: Application layer.* The final layer, the application layer, is responsible for deciding the tasks that have to be performed to allow a user to run a particular application.

2.18.2 Networks

The way in which the devices on a LAN are connected together is called the *topology*. As illustrated in Figure 2.37, there are three main LAN topologies: *bus, star,* and *ring*.

In the most straightforward LAN topology, devices are connected to a common linear transmission medium called a *bus*. All devices are attached to the bus, providing a simple, economical method of interconnecting devices. A failure in any one device will not affect the rest, but a single point of failure on the bus will cause the entire network to fail.

In a *star* LAN topology, all devices are connected to a central device that acts as a hub or a switch. Cables are not shared; each cable simply connects one device to the hub. More cable is required for this topology, which means that star networks are generally more expensive to implement. However, considerable redundancy is gained since a device failure will still only affect that device, but now a single point of network failure will also only affect the single device connected on that cable. This is a vast improvement over the bus topology in which the whole network would be interrupted by such a failure.

In a *ring* topology, devices are connected in a circular fashion so that data is passed around the network from one device to the next, until it returns to the transmitting device. Transmission is always in the same direction on the ring so that each device always receives data from the same upstream device and passes it on to the same downstream device. Each device acts as a receiver for data directed to it, and acts as a repeater for data to be passed on to downstream devices.

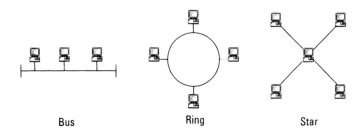

Bus   Ring   Star

Figure 2.37 Bus, ring, and star LAN topologies.

However, LAN topologies are not as simple as the above physical descriptions imply. Often a LAN technique uses a certain logical topology (i.e., the information will travel in accordance with one of the above topologies), but the LAN will be wired physically to look like another topology. The most common examples are networks that are either a logical bus or ring, but are wired to be a physical star topology.

2.18.2.1 Network Elements

A LAN will normally comprise a number of elements including a *network server* (more traditionally called a *file server*), which manages the network and the network file system and provides a number of functions such as printers, disk space, application programs, backup tape drives, and the network operating system; a *network operating system* (NOS), which runs on the server and provides the basis for resource sharing by the other devices on the network; *devices* (e.g., workstation, printers, and disk drives); and *network interface cards* (NICs), which provide the interface between the network device and the network.

2.18.2.2 LAN Media

The *LAN medium* is the physical transmission medium used to connect devices together. LAN media are traditionally cable-based although wireless links are increasingly popular.

Coaxial cable is used in Ethernet LANs, but is more commonly replaced in modern networks with twisted pair cable due to its light weight, low cost, and ease of deployment. Both UTP and shielded twisted pair (STP) are used. Fiber optic cables can be used for all elements of a LAN, including to the desktop. More commonly, however, fiber is used in the backbone and twisted pair or coaxial cable is used to connect devices. This takes advantage of both media: Optical fibers provide the high capacity for the backbone, and twisted pair provide the light weight and ease of termination required for each workstation. Wireless networks use electromagnetic radiation as the medium, either in the RF bands or the infrared band. The use of radio or infrared frequencies does not affect the LAN topology, although a different interface method is required.

2.18.2.3 Media-Access Techniques

Media-access techniques allow the devices on the LAN to gain access to the shared media. All devices cannot access the LAN simultaneously; there must be some protocol by which an orderly, timely access is available to all devices. There are two main baseband protocols employed: carrier-sense multiple-access/collision-detect (CSMA/CD), utilized by the Ethernet; and token passing, utilized by the token ring.

- *Ethernet: Carrier-sense* refers to the way in which a device that wants to transmit checks the LAN for a signal to see if any other device on the network is transmitting data. If the LAN is busy, the device will wait until the other device is finished. When no signal is present on the LAN, the device can transmit. A problem will still occur, however, if two devices simultaneously check that the network is quiet and decide to transmit data. If that happens, the data arriving at the receiving station will be corrupted. Thus, a protocol is required to prevent transmissions in those circumstances. CSMA/CD uses a *collision-detect* protocol. Collision detect requires a transmitting device to monitor the LAN while it is transmitting. If the device detects a collision, it will stop transmitting, wait a predetermined time, and then sense the LAN before attempting to retransmit.

- *Token ring:* The *token passing* media access method is more orderly than CSMA/CD. Access control is provided by a small frame of data called a *token*, which circulates around the LAN until a device wishes to transmit, in which case it takes the token and transmits a data frame. Each device on the LAN looks at the data frame to see if it is the addressee. If so, it copies the frame and allows it to pass once it has changed the last field in the frame to signify that the frame has been copied. Finally, the frame arrives back at the originating device, which can check if the receiving device has copied the frame. The originating device then releases the token to signify that the LAN is idle. At any moment, therefore, either a token or a data frame is traveling around the LAN.

The Ethernet is cheaper than the token ring but is less efficient in its use of the channel. The token ring provides an orderly access technique, while the Ethernet is far less orderly. The token ring has higher overheads at low levels of traffic but is better under a heavy workload than the Ethernet and is therefore better at transferring large files, particularly on a busy LAN. The differences are hard to quantify, although, generally, if the network will be operating at a medium workload, the Ethernet would be preferred. If a heavy workload is expected, the token ring would be preferred.

There are four ways in which these networks can be interconnected: *repeater*, *bridge*, *router*, and *gateway*. These devices operate at different levels of the OSI reference model and are also illustrated in Figure 2.38. As described in the following sections, repeaters and bridges extend networks; routers and gateways are genuine internetworking devices.

2.18.3 Network Extension

The repeater is a device that operates at layer 1 and simply retimes and regenerates the signal to transfer data from a network of one type to a network of the same type. Repeaters therefore allow cable lengths to be extended, increasing the coverage of a particular network. Media access techniques such as the Ethernet and the token ring are constrained in their length, however, and there is maximum length for a particular network that constrains those techniques to the local area.

A bridge is slightly more intelligent than a repeater and operates at layer 2 and therefore can only interconnect LANs of the same type. Bridges read the destination and source addresses in the packets or frames and forward them on to another LAN depending on the address. This simple rule means that the bridge acts as a filter, only sending frames onto a segment if they are destined for a device on that segment. On Ethernet LANs this is a very desirable feature since the segments can effectively operate as independent LANs until frames need to be sent from one to another. This reduces the potential collisions on each segment and, unlike repeaters, allows the design rules for the maximum number of segments to begin again. A bridge could therefore be used to join two maximum-length Ethernet segments.

2.18.4 Internetworking

Routers are configurable, multiport devices that forward data based on the destination address and the information contained within their configuration. The network manager can set the configuration to route traffic by different paths depending on a number of parameters such as priority, required

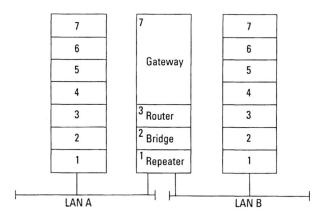

Figure 2.38 Interconnection of two networks using a repeater, bridge, router, and gateway.

response time, or cheapest path. Unlike bridges that extend networks, routers are used to interconnect LANs and to provide segmentation of a larger network rather than forming one. Routers provide similar filtering and bridging functions as bridges, but they offer more sophisticated capabilities such as network management and the ability to share network information with each other using routing protocols running at layer 3. They also offer limited security at this low level through the ability to control access lists and protocols. Bridges have no ability to share layer 3 information and therefore cannot perform the same load balancing and other network functions.

Two totally different networks can be connected by a *gateway*. A gateway operates at layer 7 (the application layer) and translates between two different protocol stacks. The gateway runs both protocol stacks—corresponding to the two stacks in the two disparate networks being connected. At layer 7, the gateway can be considered to be a translator between host machines on the two networks. In addition to protocol conversion, the gateway routes data and therefore it is by far the slowest interconnection device. If protocol conversion is required, however, this lower speed is normally accepted gladly as the cost of performing protocol conversion in each device is prohibitive and a gateway can perform the translation on behalf of a complete LAN.

2.18.5 The Internet

The best modern example of an internetwork is the Internet, which is made up of wide a variety of computer networks, including those of government departments, universities and colleges, commercial and nonprofit organizations, and individual home computers. Each of the millions of computers connected to the Internet can send or retrieve information to or from any other computer on the network, allowing users to access information from across the world. Computers can be connected to the Internet either directly through a server, or through a local network, which is then connected to the Internet through a gateway. To gain access to the Internet, individuals must first subscribe to an *Internet service provider* (ISP), which will act as the gateway to the Internet. The Internet itself is formed by the interconnection of routers, which creates a unique communications system in which there is no central point of control.

The Internet operates on the principle of packet switching. For individual access, the information is sent from the computer's modem, along the telephone line to the ISP server, where it is broken down into packets. Each packet is labeled with the address of the recipient. The nature and speed of the packets that can be sent and received depend on the memory capacity of the computers involved and the bandwidth capacity of the lines, trunks, and connections linking them.

Within the Internet, each router reads each addressed packet and sends the data in the right general direction to the next server. None of the routers has a map of the whole Internet. They only relate to their own immediate area and find the best pathway to get to the next point. When computers are on the same network, the information flow is direct, but when computers are on different networks, the information is routed through a gateway. Most gateways are dynamic in that if a packet is incorrectly forwarded to a gateway, it returns the packet. The preceding gateway then remembers where to route future packets carrying that particular destination.

Protocols are sets of rules that control transmission between components in a network. The *Transmission Control Protocol/Internet Protocol* (TCP/IP) [18], developed by the U.S. Department of Defense in 1972, is the model used on the Internet. Depending on the traffic flow, different parts of the same message may be sent by entirely different routes, only to be electronically stitched back together at the final destination. TCP/IP's role is crucial in deciding how the data will get to its destination and checking it for errors and conformity at the receiving end.

TCP provides services to applications. For example, it breaks an e-mail message down into packets, marks each packet with the information needed for reconstruction, retransmits lost datagrams, and reconstructs messages. IP provides datagram transmission services for TCP and other applications, including deciding upon the route through the network.

Every computer that connects to the Internet is given its own discreet address, which is necessary to provide a unique identification for each computer. The identification is known as an *Internet address*, or *IP address*. Computers connected to the Internet through a network will generally have a permanent IP address. A home computer, or a mobile user, will not normally be assigned a permanent IP address but, on connection, will be allocated one from a pool belonging to the user's ISP.

2.19 Types of Systems

2.19.1 Types of Transmission

Communications systems normally utilize one of three basic types of transmission: *simplex* transmission, in which a transmission path can carry information in only one direction; *half-duplex* transmission, in which a transmission path can carry information in both directions but in only one direction at a time; and *duplex* transmission, in which a transmission path can carry information in both directions simultaneously. The distinction is not always made between simplex and half-duplex, with common commercial use referring to simplex and duplex as a way of delineating whether information is being carried simultaneously on the link. In military systems, however, it is common to distinguish between the three types of transmission, particularly since half-duplex is a very common form of military transmission.

2.19.2 Satellite Communications

In satellite communications [19] the transmitted energy is directed in a line-of-sight path at a satellite, which is normally in geostationary orbit. The satellite then translates the transmission frequency and retransmits the energy, in either a wide or a narrow beam, back to receiver stations on the Earth. In essence, a communications satellite is a relay or rebroadcast station in the sky. Its function is to interconnect Earth stations so that information may pass between them. However, it is limiting to think of a satellite as nothing more than a cable in the sky linking two Earth stations. The signal received by the satellite is amplified, translated in frequency, and retransmitted to Earth over a wide area. By the use of multiple-access techniques discussed later, it is possible for a large number of ground stations to communicate on a net basis, with each transmission being received by a number of stations.

Figure 2.39 illustrates the three basic types of satellite orbit. When the satellite rotates in an orbit above the equator, it is called an *equatorial orbit*.

When the satellite rotates in an orbit that takes it over the north and south poles, it is called a *polar orbit*. Any other orbital path is called an *inclined orbit*. The altitude selected for the satellite above the Earth directly determines both the *period* of the orbit (time to complete one cycle around the Earth) and the velocity that must be imparted to the vehicle to achieve that orbit.

In addition to orbit shape, orbits are normally grouped by their altitude. *Low-Earth orbit* (LEO) satellites have altitudes of 150 to 1,500 km. *Medium Earth orbit* (MEO) satellites have altitudes of 1,500 to 35,786 km. *Geosynchronous* satellites have a period of rotation that is synchronized to that of the Earth or some multiple of it. The *geostationary* orbit is a unique geosynchronous one, located over the equator with a 0° inclination. The satellite in geostationary orbit has a height and velocity such that it appears stationary to earth-bound observation and the satellite is said to be in *geostationary Earth orbit* (GEO). The height above the Earth's surface required for geostationary orbit is 35,786 km with a velocity of 3.073 kms^{-1}. At that height, three satellites can provide worldwide coverage.

2.19.2.1   GEO

As we have seen, geostationary satellites have a rotation period of 24 hours, so that they remain in a fixed position with respect to a given Earth station and have a 24-hour availability. Once maneuvered into their "parking orbit," satellites must be carefully placed into predefined "slots," or parking places, to ensure that they, and the signals they receive and emit, do not interfere with the many other satellites nearby. When satellites are no longer usable, usually because they are running low on maneuvering fuel, they are often moved into a much higher "disposal orbit" and shut down. International agreements have defined 180 satellite slots (providing 2 degrees of separation between satellites in GEOs), but improvements in technology have now made possible the sharing by several satellites of some more desirable orbital slots, such as those over Western Europe.

GEOs are the most commonly used communications satellites. However, the high altitude of the GEOs results in round-trip transmission times of 0.5 second (0.5-second delay before each speaker hears a response). This is much longer than telephone users experience in typical terrestrial connections and can be somewhat annoying. Furthermore, the *error detecting and correction* (EDC) methods that have been developed for terrestrial systems do not operate efficiently in the presence of such long transmission times. However, special EDC methods have been developed specifically for GEOs.

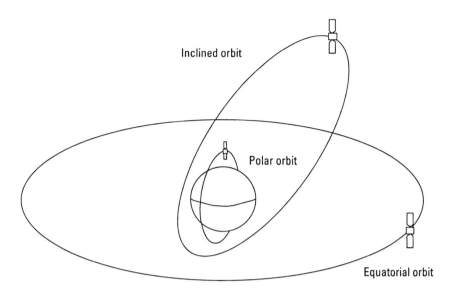

Figure 2.39 Polar, inclined, and equatorial orbits.

The advantages of GEOs are as follows:

- The GEO satellite remains almost stationary with respect to a given Earth station. Consequently, expensive tracking equipment is not required at the Earth stations, which can use simpler antennas and therefore are significantly cheaper.

- There is no need to hand over from one satellite to another as they orbit overhead. Consequently, there are no breaks in transmission because of the handover times.

- Satellites in inclined orbits require an acquisition operation and sometimes involve handover from an orbiting satellite leaving the area to a new satellite entering the area.

- High-altitude geostationary satellites can cover a much larger area of the Earth than their low-altitude orbital counterparts and global coverage can be achieved with just three satellites.

- Inclined orbits usually require multiple satellites to be spaced along the orbit in order to provide continuous coverage to a particular Earth station.

There are, however, a number of disadvantages of GEO orbits:

- Points on the Earth beyond about ±75 to 80° latitude are not visible from the satellite. Inclined orbits, however, can provide visibility to the higher northern and southern latitudes, although they require Earth stations to continually track the satellite.
- The higher altitudes of GEO satellites also introduce much longer propagation times.
- The round-trip propagation delay between two Earth stations through a GEO satellite is 500 to 600 ms, which is a bearable delay for a telephone conversation over a single satellite hop, but a two-hop telephone conversation is very difficult.
- GEO satellites also require higher transmit powers and more sensitive receivers because of the longer distances and greater path losses.

2.19.2.2   MEO

In the early days of satellite communication, the lower altitudes of MEOs (up to 10,000 km) were the only ones that could be reached with the launch vehicles available at the time. As greater payloads became possible, GEOs became used almost exclusively for communications, despite the drawback of long delay. More recently, however, MEOs have experienced a resurgence of interest because their transmission path loss is much less than GEOs, and the round-trip time is reduced to 67 to 200 ms. Circular MEOs have periods of 8 to 12 hours. As a result of the lower orbit, they do not travel at the same speed as the Earth and several MEOs are needed to provide continuous coverage.

Although GEOs seem likely to continue their domination, the lower transmission path loss of MEOs makes them particularly attractive for mobile satellite systems because hand-held terminals with much lower power and simple omnidirectional antennas can be used.

2.19.2.3   LEO

An LEO provides a further reduction in transmission path loss and in round-trip delay (on the order of 13 ms). This allows the use of even lower-power, handheld terminals in *mobile satellite service* (MSS) systems. The altitude of an LEO should be high enough to avoid substantial deceleration due to the atmospheric friction, but low enough to avoid the more intensive levels of proton bombardment in the inner Van Allen belt. For communications, altitudes between 780 and 1,400 km are favored, corresponding with orbital periods between 100 and 113 minutes. Thus, LEO systems require slightly more satellites than MEO systems to provide continuous coverage. For example, the Iridium system uses 66 satellites.

2.19.2.4   Advantages and Disadvantages of LEO/MEO Orbits

The lower-orbit satellites have the following advantages:

- Less transmitter power is required.
- There is minimal propagation delay.
- Smaller antennas are required.
- LEOs are less subject to shadowing than GEOs.

However, LEO/MEO satellites have the following disadvantages:

- Larger numbers of satellites are required,
- There is an increased requirement for handover.
- Satellites tend to have a shorter lifetime due to orbital decay.

2.19.3   Personal Communications System

A personal communications system (PCS) provides communications services to mobile terminals. PCS operates in the UHF band, with frequencies typically lying between 800 MHz and 3 GHz. All communications in a PCS occur between a mobile station and one of a number of base stations, which are in a fixed location. Each base station is connected to a fixed communications network that provides switching between mobile stations and connection to the public telephone network. In fact, it should be remembered that the subscriber is the only mobile element in a mobile communications system. While PCS allows great mobility for subscribers, that mobility comes at some considerable infrastructure cost for the network provider.

The basic elements of a PCS system are shown in Figure 2.40. The mobile subscriber units may be mounted in a vehicle or carried as a portable. Subscribers are assigned a duplex channel with which to communicate to a designated base station, which communicates simultaneously with all mobiles within its coverage area (called a *cell*). Base stations are connected to mobile switching centers, each of which controls a number of cells and handles connections with the fixed PSTN. Each base station is allocated a different carrier frequency with which to communicate with subscribers. Because only a limited portion of the radio spectrum is available for cellular systems, the number of carrier frequencies available is limited. This means that the available frequencies must be reused many times in order to provide sufficient channels to cope with a large number of subscribers.

Each base station has allocated to it a number of channels that can be used for voice communications or for control and signaling traffic. When a mobile is active, it registers with an appropriate base station and the mobile's cellular location is stored in the mobile switching center responsible for that base station. When a call is established either from or to the mobile, the control and signaling system instructs the mobile to use one of the base-station channels through which it can obtain access to the fixed network. When the call is completed, the mobile releases the voice channel that can then be reallocated to other users.

As the mobile moves around the cell, the quality of its allocated channel will vary as the signal path varies and as the mobile gets further away from the base station. Throughout the call, the quality of the radio channel is monitored so that if it falls below a set threshold (most probably because the mobile is about to leave the cell), the mobile can be instructed to *hand over* (or *hand off*) to the base station that has the strongest signal. In this way, a continuous service can be provided to mobiles as they travel around the network.

The capacity of a base station is provided by sharing the carrier frequency between communications channels on a frequency division basis (typically in first generation analog systems), on a time division basis (typically in second-generation digital systems) or on a code division basis (also typical of digital systems). Most modern systems employ a combination of access techniques.

Most services of current PCS are based on circuit switching. Maximum data capacities per circuit are between 2.4 and 9.6 Kbps. There are four classes of PCS:

- Cordless communications systems;
- Terrestrial cellular communications systems [20];
- Satellite cellular communications systems, such as Iridium, Globalstar, and Teledesic [21];
- Trunked radio (also known as private mobile radio), such as TETRA [22] and APCO-25 [23], both of which are discussed in more detail in Chapter 7.

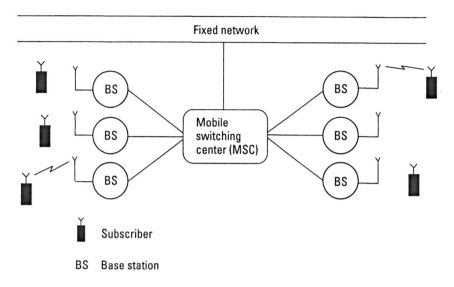

Figure 2.40 Basic elements of a cellular system.

The major difference between cordless and cellular systems is that cellular systems provide more support for mobility, including handoff of a mobile station from one base station to another.

It is currently possible to purchase a user terminal that can access more than one type of PCS, although handoff between different types of PCSs is not supported. Third-generation cellular systems [24] will go further and provide a single logical network, with handoff between satellite and terrestrial systems being supported.

Endnotes

[1] This figure is based on the model presented in Sklar, B., *Digital Communications: Fundamentals and Applications*, Englewood Cliffs, NJ: Prentice Hall, 2001, p. 5.

[2] There are a number of excellent sources for those wishing to examine further the aspects of communications technology that are presented in this chapter:

Haykin, S., *Communication Systems*, New York: Wiley, 2001.

Roddy, D., and J. Coolen, *Electronic Communications*, Englewood Cliffs, NJ: Prentice Hall, 1995.

Sklar, B., *Digital Communications: Fundamentals and Applications*, Englewood Cliffs, NJ: Prentice Hall, 2001.

Tomasi, W., *Electronic Communications Systems: Fundamentals Through Advances*, Englewood Cliffs, NJ: Prentice Hall, 2001.

Young, P., *Electronic Communication Techniques*, Columbus, OH: Merrill Publishing Company, 1990.

Ziemer, R., and R. Peterson, *Introduction to Digital Communications*, Englewood Cliffs, NJ: Prentice Hall, 2001.

[3] Further details on source coding can be found in [2] as well as:

Luther, A., *Principles of Digital Audio and Video*, Dedham, MA: Artech House, 1985.

Salomon, D., *Data Compression*, New York: Springer-Verlag, 2000;

[4] Sources of information on encryption techniques include:

Denning, D. E., *Cryptography and Data Security*, Reading, MA: Addison-Wesley, 1983.

Schneier, B., *Applied Cryptography: Protocols, Algorithms, and Source Code in C*, New York: Wiley, 1994.

Singh, S., *The Code Book: The Science of Secrecy from Ancient Egypt to Quantum Cryptography*, New York: Anchor Books, 1999.

Sinkov, A., *Elementary Cryptanalysis: A Mathematical Approach*, New York: The Mathematical Association of America, 1966.

Stallings, S., *Network and Internetwork Security*, 2nd ed., Englewood Cliffs, NJ: Prentice Hall, 1995.

Torrieri, D. J., *Principles of Secure Communications*, Dedham, MA: Artech House, 1985.

[5] Further details on channel coding can be found in [2] as well as:

Oberg, T., *Modulation, Detection, and Coding*, New York: Wiley, 2001.

Purser, M., *Introduction to Error-Correcting Codes*, Norwood, MA: Artech House, 1995.

[6] Frater, M., and M. Ryan, *Electronic Warfare for the Digitized Battlefield*, Norwood, MA: Artech House, 2001.

[7] For more details, see: Shakhgildyan, V., *Radio Transmitter Design*, Moscow: MIR Publishers, 1987.

[8] For more details, see: Rohde, U., and T. Bucher, *Communications Receivers: Principles and Design*, New York: McGraw-Hill, 1988.

[9] For more details, see: Hoss, R., *Fiber Optic Communications: Design Handbook*, Englewood Cliffs, NJ: Prentice Hall, 1990.

[10] For more details, see: Collin, R., *Antennas and Radiowave Propagation*, New York: McGraw-Hill, 1985.

[11] For more details, see: Rohan, P., *Introduction to Electromagnetic Wave Propagation*, Norwood, MA: Artech House, 1991.

[12] For more details, see: Maclean, T., and Z. Wu, *Radiowave Propagation over Ground*, London: Chapman & Hall, 1993; and Collin, R., *Antennas and Radiowave Propagation*, New York: McGraw-Hill, 1985.

[13] Further details on sky-wave communications can be found in:

Betts, J., *High Frequency Communications*, London: The English Universities Press Ltd, 1967.

Braun, G., *Planning and Engineering of Shortwave Links*, Berlin: Siemens Aktiengesellschaft, Wiley, 1986.

Goodman, J., *HF Communications: Science and Technology*, New York: Van Nostrand Reinhold, 1992.

Maslin, N., *HF Communications*, London: Pitman, 1987.

[14] For more details, see: Schanker, J., *Meteor Burst Communications*, Norwood, MA: Artech House, 1990.

[15] For more details, see: Roda, G., *Tropospheric Scatter Radio Relay Links: Guide to Design and Implementation*, Castallanza, Italy: Applicazioni Radio Elettroniche SPA, 1986.

[16] Black, U., *Internet Architecture: An Introduction to IP Protocols*, Upper Saddle River, NJ: Prentice Hall, 2000.

[17] Further details on TCP/IP can be found in [2] as well as:

De Prycker, M., *Asynchronous Transfer Mode: Solution for Broadband ISDN*, New York: Ellis Horwood, 1993.

Goralski, W., *Introduction to ATM Networking*, New York: McGraw-Hill, 1995.

[18] Further details on TCP/IP can be found in [2] as well as:

Stevens, W., *TCP/IP Illustrated*, Reading, MA: Addison-Wesley, 1996.

Wilder, F., *A Guide to the TCP/IP Protocol Suite*, Norwood, MA: Artech House, 1998.

[19] Further details on satellite communications can be found in:

Elbert, B., *Introduction to Satellite Communication*, Norwood, MA: Artech House, 1999.

Elbert, B., *The Satellite Communication Applications Handbook*, Norwood, MA: Artech House, 1997.

Gagliardi, R., *Satellite Communications*, New York: Van Nostrand Reinhold, 1984.

Ha, T., *Digital Satellite Communications*, New York: Macmillan Publishing Company, 1986.

Jansky, D., *Communications Satellites in the Geostationary Orbit*, Dedham, MA: Artech House, 1983.

Kadish, J., and T. East, *Satellite Communications Fundamentals*, Norwood, MA: Artech House, 2000.

Maral, G., and M. Bousquet, *Satellite Communications Systems*, 3rd ed., New York: Wiley, 1998.

Martinez, L., *Communication Satellites: Power Politics in Space*, Dedham, MA: Artech House, 1985.

Matos-Gomez, J., *Satellite Broadcast Systems Engineering*, Norwood, MA: Artech House, 2002.

Pratt, T., and C. Bostian, *Satellite Communications*, New York: Wiley, 1986.

Pritchard, W., H. Suyderhoud, and R. Nelson, *Satellite Communication Systems Engineering*, Englewood Cliffs, NJ: Prentice Hall, 1993.

Richharia, M., *Satellite Communications Systems: Design Principles*, New York: McGraw-Hill, 1995.

Roddy, D., *Satellite Communications*, New York: McGraw-Hill, 1996.

[20] Further details on mobile communications systems can be found in:

Doble, J., *Introduction to Radio Propagation for Fixed and Mobile Communications*, Norwood, MA: Artech House, 1996.

Garg, V., and J. Wilkes, *Wireless and Personal Communications Systems*, Englewood Cliffs, NJ: Prentice Hall, 1996.

Hess, G., *Land-Mobile Radio System Engineering*, Norwood, MA: Artech House, 1993.

Hernando, J., and F. Perez-Fontan, *Introduction to Mobile Communications Engineering*, Norwood, MA: Artech House, 1999.

Lee, W., *Mobile Cellular Telecommunications Systems*, New York: McGraw-Hill, 1990.

Rappaport, T., *Wireless Communications*, Englewood Cliffs, NJ: Prentice Hall, 1996.

[21] Further details on satellite-based mobile communications can be found in:

Jamalipour, A., *Low Earth Orbital Satellites for Personal Communication Networks*, Norwood, MA: Artech House, 1998.

Ohmori, S., H. Wakana, and S. Kawase, *Mobile Satellite Communications*, Norwood, MA: Artech House, 1998.

Pattan, B., *Satellite-Based Cellular Communication*, New York: McGraw-Hill, 1996.

Ryan, M., "Satellite-Based Mobile Communications," in L. Godara, (ed.), *Handbook of Antennas in Wireless Communications*, Boca Raton, FL: CRC Press, 2002.

[22] TETRA specifications are contained in:

ETS 300 392-1, "Radio Equipment and Systems (RES); Trans-European Trunked Radio (TETRA); Voice plus Data (V+D); Part 1: General Network Design," Sophia Antipolis: ETSI, 1996.

ETS 300 392-2, "Radio Equipment and Systems (RES); Trans-European Trunked Radio (TETRA); Voice plus Data (V+D); Part 2: Air Interface (AI)," Sophia Antipolis: ETSI, 1996.

ETR 300-1, "Radio Equipment and Systems (RES); Trans-European Trunked Radio (TETRA); Voice plus Data (V+D); Designers Guide; Part 1: Overview, Technical Description and Radio Aspects," Sophia Antipolis: ETSI, 1997.

ETR 300-2, "Radio Equipment and Systems (RES); Trans-European Trunked Radio (TETRA); Voice plus Data (V+D); Designers Guide; Part 2: Traffic Aspects," Sophia Antipolis: ETSI, 1997.

ETR 300-3, "Radio Equipment and Systems (RES); Trans-European Trunked Radio (TETRA); Voice plus Data (V+D); Designers Guide; Part 3: Direct Mode Operation (DMO)," Sophia Antipolis: ETSI, 1997.

ETS 300 393-1, "Radio Equipment and Systems (RES); Trans-European Trunked Radio (TETRA); Packet Data Optimized (PDO); Part 1: General Network Design," Sophia Antipolis: ETSI, 1998.

[23] Standards from APCO Project 25 are published by the Telecommunications Industry Association in their 102-series.

[24] Further details on third-generation mobile communications can be found in:

Korhonen, J., *Introduction to 3G Mobile Communications*, Norwood, MA: Artech House, 2001.

Prasad, R., W. Mohr, and W. Konhauser, (eds.), *Third Generation Mobile Communication Systems*, Norwood, MA: Artech House, 2000.

Prasad, R., (ed.), *Towards a Global 3G System: Advanced Mobile Communications in Europe, Volume 1*, Norwood, MA: Artech House, 2001.

3

Introduction to Land Force Structures

3.1 Introduction

Effective communications is vital to the command and control of any army. Many characteristics of tactical communications systems are determined by the structure of the forces that they support. It is therefore useful to examine the general structure of land forces as a precursor to analyzing the communications systems that support them. Since we are interested in tactical communications systems, we focus on the structure of land forces at the tactical level.

This chapter provides an overview of army organizational structure, aiming to provide a basic understanding of the tasks that armies are called upon to perform, the size and disposition of army units, the army command structure, tactical communications structure, and the role of support and services. The basic army structure discussed here applies to most of the world's armies. The coverage is general; we do not discuss details of any particular army.

3.2 Spectrum of Military Operations

While waging war is often seen as the principal task of an army, in practice an army's tasks are much more varied and include a range of protective and humanitarian activities that the civilian agencies of government (such as police and emergency services) are often unable to undertake. Peacekeeping, often under a United Nations mandate, is one such example. While some commentators see this as a post-Cold War phenomenon, there is in fact nothing new about this variety of tasks for military forces. The armies (and even navies) of European colonial powers, for example, were used for centuries to suppress local uprisings by natives of various colonies, or as guards for prisoners in penal colonies (e.g., in Britain's American, and later Australian, colonies).

The breadth of tasks that can be undertaken by an army is often referred to as the *spectrum of operations*, which would typically include large-scale combat operations; lower-level combat operations (often designed to prevent higher-intensity conflict), such as peace enforcement, counterinsurgency, and counterterrorism; and peacekeeping and humanitarian operations. In most countries, troops are also used to support the civil government in emergencies. These support tasks might include disaster relief (following flood, fire, earthquake, or another natural disaster), provision of essential services during strikes, and riot control. Sometimes, particular parts of an army may have special responsibilities in support of the civil government, as is the case for the U.S. National Guard.

JV2020 divides military operations into two classes: combat (i.e., war) and noncombat [also known as military operations other than war (MOOW)], incorporating all the other roles of a military force [1].

As indicated in Table 3.1, the boundaries between different types of operations are often blurred and an army must often be able to transition quickly from one different type of operation to another. A peacekeeping operation, for example, may require that a force maintain a high level of visibility as a sign of good faith and a deterrent to potential opponents. If attacked, however, the force must be able to react quickly, possibly taking action more typical of combat operations.

Many characteristics of these operations differ. One of the most obvious is the rules of engagement (i.e., the rules dictating the circumstances under which it is acceptable to use particular levels of force). Generally, the rules of engagement are more restrictive for operations listed in the lower part of Table 3.1 than for those listed in the upper part. In recent NATO air operations in Kosovo, for example, attacks against individual targets were sometimes required to be authorized at the political level in up to 19 different countries. Even within the NATO military organization, it is reported that the air commander (General Short) personally oversaw the search for three tanks [2], and in many cases was required to personally authorize attacks [3]. This type of restriction would not normally be applied in a higher-intensity conflict (e.g., the Gulf War).

Similarly, the density of troops in the area of operations also varies with the type of operation, once again being lower for operations in the lower part of Table 3.1 than those in the upper part. This is driven by a trade-off between the cost of an operation and the level of threat; where the level of threat is low (i.e., for operations listed toward the bottom part of Table 3.1), the cost of an operation can be reduced without introducing unacceptable risk by using a smaller size of force than would be used for a higher threat operation.

At the tactical level, any army is required to be able to undertake a number of tasks. These include patrolling, defending ground, protecting vital facilities, and capturing either ground or facilities.

3.3 Army Structure

Key differences between the capabilities of an army in carrying out the various types of operation and those of civil agencies (such as the civil police) are the power of the weapons available, the level of force that may be considered acceptable, and the availability of a high level of integral support. One of the key characteristics of an army that separates it from other types of organization is that it can deploy to a remote location and support itself in a hostile

Table 3.1
Range of Military Operations Outlined in JV2020

Type of Operation	General U.S. Goals	Examples
Combat	Fight and win	Large-scale combat operations
		Attack/defend/blockade
Noncombat	Deter war and resolve conflict	Peace enforcement
		Counterterrorism
		Show of force/raid/strike
		Peacekeeping
		Counterinsurgency
	Promote peace and support U.S. civil authorities	Freedom of navigation
		Counterdrug
		Humanitarian assistance
		Protection of shipping
		U.S. civil support

environment. This support typically includes providing its own communications, being able to construct roads and bridges, and maintaining the supply of food, water, ammunition, and equipment to troops.

Firepower is provided by infantry (foot soldiers), armor (such as tanks), aviation (such as helicopter gunships), and artillery. These units cannot operate without support, including communications and electronic warfare, provided by signals units; construction and demolition support (such as bridges), provided by engineers; information on adversary intentions and actions, provided by intelligence assets; transport, provided by transport units as well as armored and aviation units; supplies, provided by various supply units; and medical support, provided by field ambulance and field hospitals.

Traditionally, the functions of land forces are broken into the following areas: *combat* (infantry and armor), *combat support* (signals, engineers, artillery, and aviation), and *services* (transport, medical, and support). Here, we are concerned primarily with infantry units and the signals support that is integral to them or provided externally.

Armies are typically organized into strongly hierarchical structures, similar to that shown in Table 3.2.

The symbols shown in Figure 3.1 are often used to represent the size and type of different units. Infantry units are denoted by a box with a cross inside. Signals units are denoted by a box with a diagonal zig-zag line. Other types of units have symbols of their own. The size of a unit is indicated by the symbol above the box. A headquarters is distinguished from its unit by the addition of a mast to the flag symbol.

Table 3.2
Typical Army Units

	Size (Personnel)	Maximum Frontage in Defense
Squad/section	6–9	200m
Platoon/troop	30	500m
Company/squadron	100	1.2 km
Battalion/regiment	600–800	4 km
Brigade	3,000	12 km
Division	10,000	25 km
Corps	30,000	50 km

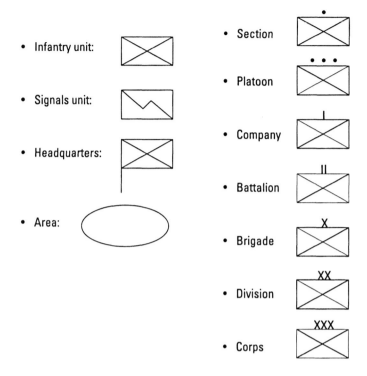

Figure 3.1 Symbols representing army units of various sizes.

This section describes the characteristics of the various levels of an army's hierarchy in conventional high-intensity combat operations. The differences in operating characteristics for other types of operation are discussed in Section 3.3.4.

3.3.1 The Infantry Squad

3.3.1.1 Organization

The infantry squad [4] (or rifle section) consists of approximately six to nine members, usually commanded by a sergeant (U.S. army) or corporal (British army). Its weapons normally include at least one light machine gun and might include rocket-propelled grenades or antitank weapons. The remaining soldiers in the squad are equipped with automatic rifles. In most armies, the infantry squad is organized internally into two or three smaller groups. In the U.S. army, for example, the squad is divided into two *fire teams* of four members. As described later in Section 3.3.4, the size of squads may decrease when infantry units have integral transport.

Figure 3.2 shows a representation of an infantry squad in defense. The ellipse represents the area occupied by the squad. The dot in the boundary indicates that the area is occupied by a squad-sized force. The circles inside represent notional locations of weapons pits that are used to provide protection from an adversary's fire. The area covered by a infantry squad when patrolling or in defense is determined by a number of factors. Vulnerability to an adversary's indirect firepower (such as mortars and artillery) is minimized by maximizing the spacing between squad members. The maximum separation is limited by tactics, communications, and logistics. Tactical considerations require mutual support between the squad's weapons and the requirement to provide all-around defense. The terrain also has a significant impact on the area: It is usually required that each soldier can see those other soldiers closest to him or her. The range of communications also limits separation within the squad. If communications is by voice, then the separation is limited to shouting distance. Finally, the ability to move supplies around the squad area (usually on foot) limits the separation.

3.3.1.2 Communications

Communications within a squad is usually face-to-face, either by voice or hand signals, or by hand carriage (runners). Squad radio (UHF) is sometimes

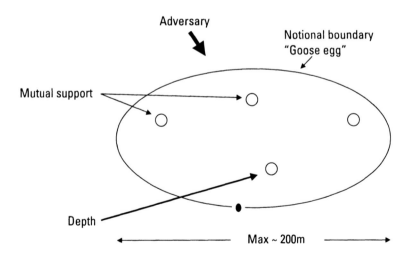

Figure 3.2 Infantry squad in defense.

used. Communications cord is sometimes used in defense to relay very simple messages.

3.3.2 The Rifle Platoon

3.3.2.1 Organization

A rifle platoon consists of a headquarters, usually containing a platoon commander of lieutenant rank and a second-in-command (2IC) of sergeant rank, possibly a support section supplying additional (heavier) weapons, and three or four infantry squads.

Figure 3.3 illustrates a platoon in defense. The ellipse indicates the area occupied by the platoon. The three dots in the boundary indicate that the area is occupied by a platoon-sized force. Squads are laid out to provide mutual support, all-around defense, and depth. The platoon headquarters is positioned to provide effective command and control of the platoon. As was the case for the squad, maximum protection from an adversary's indirect fire is obtained by maximizing the area over which the platoon is deployed. This maximum area is limited by tactics, communications, and logistics. Larger separations are possible in the platoon because mutual support between squads is provided by the squad machine guns, as opposed to mutual support within the squad, which is provided by the soldier's personal weapons. Like the squad, the platoon does not have any integral support assets and relies heavily on higher levels for these.

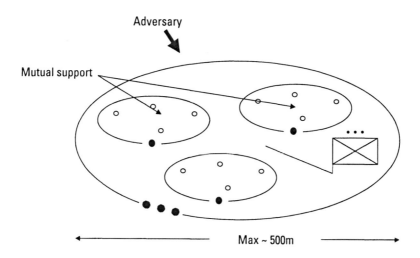

Figure 3.3 Rifle platoon in defense.

3.3.2.2 Communications

Communications within a rifle platoon employ face-to-face communications and hand carriage (runners). VHF radio or point-point telephone means are sometimes used for communications between platoon headquarters (HQ) and squads.

3.3.3   The Rifle Company

3.3.3.1 Organization

A rifle company usually consists of [5]:

- A headquarters, usually containing a company commander (with the rank of major or captain) and 2IC and company sergeant major;

- Probably a support section supplying additional weapons, especially if platoons do not have such a section;

- Three or four rifle platoons.

Like lower levels, the company does not have integral support assets. It therefore relies heavily on higher levels. A rifle company in defense is represented in Figure 3.4. The structure is the similar to that of the platoon, still employing the basic principles of depth and mutual support. Each of the platoons is deployed in accordance with these same principles, as outlined above. Once again, separation between elements of the company is key to defense, but is limited by tactics, communications, and logistics. Tactically, the company is able to deploy over a wider area than a platoon or squad because of the availability of heavier weapons from the support section to provide mutual support between platoons.

3.3.3.2 Communications

Company communications use the following means: face-to-face, VHF radio, line (telephone), and hand-carriage (runners). In most armies, the company is equipped with radio to provide communications for the command and control of its platoons. Each company has one radio net, with a primary and at least one alternate frequency, and approximately 10 RF emitters. These numbers are multiplied by three to four if a platoon radio net is used.

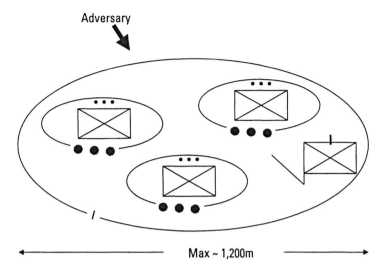

Max ~ 1,200m

Figure 3.4 Rifle company in defense.

3.3.4 The Infantry Battalion

3.3.4.1 Organization

The battalion differs from lower levels both in size and in the range of integral support assets. This increased range of integral support provides greatly enhanced flexibility. A typical infantry battalion consists of [6]:

- Battalion headquarters, including a commanding officer (CO) of lieutenant-colonel rank, 2IC (major rank), and an operations officer (major rank);

- Three or four rifle companies;

- Support assets of platoon strength, often including mortars, engineer support (sometimes known as assault pioneers), machine guns, signals, antiarmor, transport, technical maintenance, and medical.

Figure 3.5 shows an infantry battalion in defense. The boundaries are shown as straight lines for illustrative convenience; in practice they would usually follow terrain features such as roads, rivers, or contour lines. Each boundary is marked with a symbol indicating the level of the command hierarchy associated with the boundary. The battalion headquarters would usually be

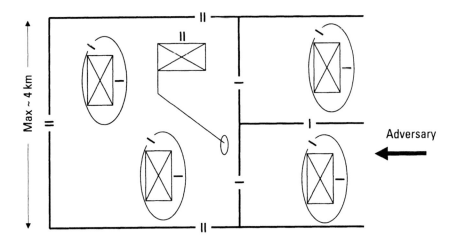

Figure 3.5 Infantry battalion in defense.

deployed somewhere to the rear of the battalion area. Sometimes, the commander may deploy an additional tactical HQ further forward.

Because of its integral support, the infantry battalion is much more capable than a group of rifle companies. These support assets provide it with a range of weapons not usually available at lower levels (e.g., mortars and heavier machine guns), communications, and engineer support, as well as a range of logistics assets that enable the battalion to sustain itself for a short period of time.

3.3.4.2 Communications

Communications for command and control within the battalion is provided by the battalion signals platoon, if present. The battalion is therefore the lowest tactical level at which specialist communicators are available. Battalion communications is based primarily on VHF radio and simple telephony. VHF radio is used where possible, with HF being used for subunits employed on tasks beyond the range of VHF, as might be the case on some reconnaissance and patrolling tasks. UHF radio may be used for ground-air communications. Traditionally, the battalion has only a manual telephone switchboard, which is not connected to outside communications. More recently, it has been usual for the battalion commander to have a single direct connection into the higher-level trunk network.

Each battalion has a total of up to 10 radio nets, each with its own primary and alternate frequencies, and a total of up to 100 RF emitters.

3.3.5 The Infantry Brigade

3.3.5.1 Organization

An infantry brigade [7] consists of a headquarters, a signals squadron, and attached units. Often, these attached units are actually owned by the division and allocated to the brigade for particular tasks. The brigade headquarters typically includes the commander, an operations command post [ops CP, or tactical operations center (TOC)], a personnel and logistics command post (pers/log CP or TOC), and a defense platoon. The brigade signals the squadron that provides the communications requirements for the brigade, including communications to the units under command. Attached units would usually include three or four infantry battalions, a regiment of armored personnel carriers (APCs), an artillery regiment, an engineer regiment, and a supply squadron and other logistics units.

Figure 3.6 shows a brigade in defense. At the coarse level of detail shown, the structure is similar to the battalion. One important addition is the brigade maintenance area (BMA), also known as the brigade support area (BSA) in the rear. This maintenance area is used by the brigade for moving

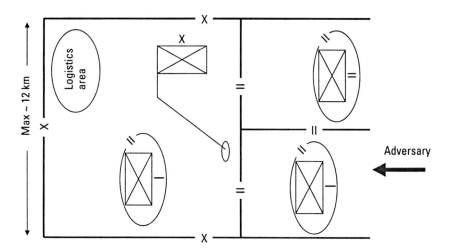

Figure 3.6 Brigade in defense.

stores and reinforcements through to its units. It is also used by member units for various administrative functions.

3.3.5.2 Communications

Brigade communications are provided by the signals squadron, which consists of specialist communicators whose primary role is to support the brigade commander's command and control of the brigade. In some armies (such as the British), the brigade signals squadron is also responsible for providing administrative support to the brigade headquarters.

Brigade communications use VHF and HF radio, as well as UHF for ground-air communications, line (extensive telephone systems with automatic switchboards within headquarters), trunk communications linking the brigade HQ telephone system to higher HQ and possibly to battalion commanders), and hand carriage (vehicle-based).

Each brigade (including attached divisional units) has approximately 50 radio nets, and 600 RF emitters. These numbers assume radio is used only at the company level and above. In areas where radio is used at the platoon level, the number of nets and emitters would be multiplied by a factor of three to four.

3.3.6   The Infantry Division

3.3.6.1 Organization

The infantry division is a formation that includes the support units required to enable sustained operations without external support [8]. A typical division has a headquarters with similar structure to a brigade headquarters. This headquarters is often split into two parts, known as *main* and *rear*. Its communications are provided by a signals regiment. A division usually possesses sufficient units of various types to support three brigades, each with a full complement of infantry, armor, engineers, artillery, and supply.

In most traditional armies, the division is the lowest level formation with sufficient integral support and services to conduct sustained operations without additional external support units. This ability to sustain itself gives a division capability well beyond that of three brigades. Figure 3.7 represents a division in defense. The structure is very similar to a brigade and includes a large divisional maintenance area (DMA), also known as the division support area (DSA), housing logistics facilities, in the rear.

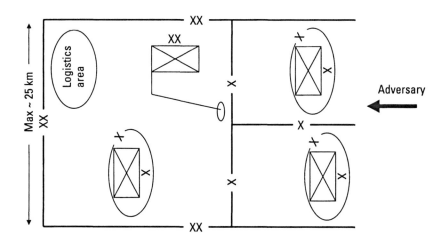

Figure 3.7 Division in defense.

3.3.6.2 Communications

The communications within a division are provided by its signals regiment [9]. The majority of division-level communications are handled by trunk systems, providing a range of high-capacity communications services. This is due to the very large volume of data and telephony that is required by a modern division, especially for logistics. Single channel radio communications is also provided (both VHF and HF) and primarily used for command.

Each division has approximately 160 radio nets and 2,000 RF emitters. These numbers assume that radio is used only at the company level and above. In areas where radio is used at the platoon level, the number of nets and emitters would be multiplied by a factor of three to four.

3.3.7 The Infantry Corps

3.3.7.1 Organization

The infantry corps consists of a headquarters, two signals regiments, three infantry divisions, and a number of corps assets such as an EW regiment. Figure 3.8 represents a corps in defense. Like the division and brigade, the corps sets aside an area in the rear for its logistics, known as the *corps maintenance area* (CMA) or *corps support area* (CSA).

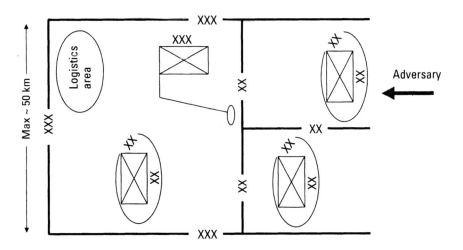

Figure 3.8 Corps in defense.

3.3.7.2   Communications

Corps-level communications rely almost exclusively on high-capacity trunk communications systems; single channel combat radio would provide only a backup means. Large amounts of data would often be transferred, requiring high-bandwidth communications systems. Long range may also be required, especially in communicating with higher headquarters.

Each corps has approximately 500 radio nets and 6,000 RF emitters. These numbers assume radio is used only at the company level and above. In areas where radio is used at the platoon level, the number of radios and radio nets would be multiplied by a factor of three or four.

3.3.8   Mechanized and Motorized Infantry

In modern warfare, speed of maneuver is critical. On foot, infantry can move only approximately 4 km/hr. Higher speed of maneuver can be obtained by providing transport in the form of trucks or APCs. Flexibility of maneuver is maximized by making these transport assets integral to units and formations.

The term *motorized infantry* refers to infantry units with integral truck transport of sufficient capacity to uplift the whole unit. *Mechanized infantry* refers to infantry units with integral armored transport (APCs) with sufficient capacity to move the whole unit [10].

The size of an infantry squad in mounted infantry units may be smaller than described above. One possible reason for this is the provision of fire support provided by a gun mounted on the vehicle.

For both mechanized and motorized infantry, it is usual that each troop-carrying vehicle would carry a radio. This results in a multiplication of the number of RF emitters in a given unit of formation by a factor of approximately two to three.

3.3.9 Span of Command

The span of command of the typical army is illustrated in Figure 3.9. This hierarchical structure, with a span of command of approximately three at each level, is a product of thousands of years of military experience. In practice, commanders at most levels also have integral combat support and services, making their effective span of command up to approximately seven, which is approximately the typical capacity of human short-term memory [11]. This structure also provides flexibility in the number of different-sized groupings that are available for tasking.

Compared to civilian organizations, the number of levels of hierarchy in a military organization appears large, and the span-of-command low. The apparent difference between military and civilian organizations is exaggerated by an eagerness by management in civilian organizations to see structures as flatter than they really are. The lowest level of manager in a civilian organization may, for example, be directly responsible for a group of 20 to 30

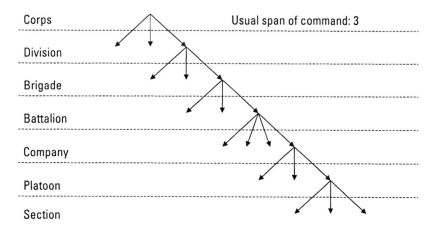

Figure 3.9 Typical army command structure, excluding combat support and services.

people. It is common, however, for this manager to be supported by two or three individuals with the title of "team leader," each of whom is responsible to the manager for supervising a portion of the group. In other words, while there may be no formal equivalent of the squad in this civilian hierarchy, in reality there are individuals who are assigned the role of the squad commander.

Another important difference between military and commercial organizations is the need for flexibility. The crew members of a commercial airliner, for example, are carefully trained to carry out all tasks in a particular way and set out in detail in the airlines' operational procedures. This maximizes the flexibility of crew rosters, allowing a crew to be formed from any group of qualified individuals, and removing the need for crew members to train together before they work together. A military organization trained in this way would be very predictable in its operations, making it extremely vulnerable against any adversary that understood its procedures. It is more usual, therefore, for an army to adopt doctrine that sets out the general principles by which it operates, but retain significant flexibility at all levels of the command hierarchy. This approach requires, however, that troops train in the same groups in which they will operate, and provides a large range of modular units that can be allocated to tasks of different sizes. A task appropriate for a company-sized force cannot normally be performed by three platoons without the addition of a company headquarters and the organization training as a company. Removal of one or more levels of the command hierarchy can be seen, therefore, to reduce significantly the flexibility of a force.

3.4 Joint Forces

Historically, the different branches of the armed services (navy, army, air force, and, in some countries, marines) have operated largely independently of one another at the tactical level. Coordination of support between the services, such as naval gunfire support or close air support for land forces, has therefore been carried out at a very high level. This is seen as placing unnecessary barriers to the provision of support between the services and reducing the effectiveness of the support that is provided, including reducing the combat power that can be brought to bear on an adversary.

Integration of elements from all three services into a *joint force* at the tactical level is seen as an important enabler for future operations, both high and low intensity, combat and noncombat. Such a joint force would consist of elements from two or more services, but not be confined to traditional force structures that assign particular capabilities (such as fire support) to forces of a particular size. For example, a joint force might include only one battalion of infantry, but be assigned naval gunfire support and a land-based fire support capability traditionally associated with a division.

Integration into a joint force implies a certain level of integral support within each element of the force. For the land element, this typically does not exist at levels below battalion, suggesting that direct integration of lower levels (such as company) into a joint force would not commonly occur. Issues associated with the training of commanders for joint operations also suggest that integration below this level is likely to be difficult. It may be that even battalion is too low a level for many types of operation, with integration occurring only at the formation (i.e., brigade or division) level.

Because the issues of span of command discussed in Section 3.3.9 are related primarily to the capabilities of human commanders and their human staffs, the fundamental principles of span of command should not be expected to change as a result of the use of joint forces.

3.5 Dispersed Forces

The high density of forces described in Section 3.3.3 is necessary in current-generation armies for conventional, high-intensity conflict (i.e., the operations at the top of Table 3.1). The dispersal of a force over a larger area creates distances between its various elements too great for them to provide mutual support, thereby increasing their vulnerability. In operations where the level of threat is lower, such as peacekeeping or low-level conflict, it is often desirable to disperse a force over much larger areas, often as a means of cost reduction. When operational requirements force dispersal of forces over large areas, defense of the whole area is not feasible. In defense, units and formations rely on intelligence and reconnaissance data to position its forces at critical locations. This might include protection of key assets, such as the civil telecommunications infrastructure. Figure 3.10 illustrates the layout of a widely dispersed brigade, in which the three battalions under command are spread over an area with diameter of 1,000 km. Within this, one of the battalions is itself dispersed over a wide area in three company locations.

A comparison of deployments in the Gulf War and for peacekeeping in East Timor illustrates the difference between the dispersal of forces in combat and peacekeeping operations. In the Gulf War, four corps were deployed

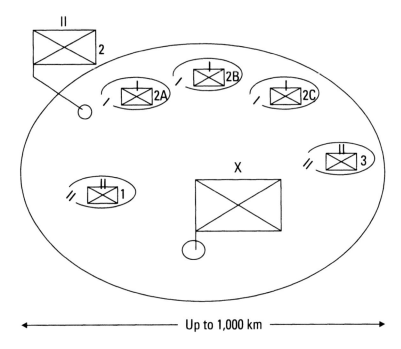

Figure 3.10 Layout of widely dispersed forces.

across a front of approximately 300 km on the northern border of Saudi Arabia. This is roughly in line with the maximum expected for high-intensity operations, allowing for each corps having a frontage of approximately 50 km, and with gaps of several kilometers between corps. In East Timor, however, a single Australian battalion was deployed on the border between East Timor and West Timor with the task of preventing infiltration of armed militias, across a front of approximately 70 km.

Increasing the distances between the various elements of a force has important consequences for all types of support, including engineers, logistics, and communications. For engineers, dispersal increases the amount of road requiring construction and maintenance. For logistics, dispersal increases traveling times between units, thereby increasing the transport assets required to service a particular force. For communications, especially those operating at VHF and higher frequencies that are limited to line-of-sight operation, dispersal increases the requirement to provide relays to extend range. The use of relays itself increases the dispersal of signals units, creating another source of additional logistics load. The impact of altering the range of communications is discussed in Section 3.6.

3.6   Characteristic Distances for Communications

One of the important ways in which the army structure influences the characteristics of military communications is through the distances involved. Table 3.3 shows characteristics distances for communications within the various levels of the army hierarchy.

The range over which a force must communicate is strongly related to the means of communications used, as illustrated in Figure 3.11. At the lowest levels, the primary means of communications is face-to-face, and the required range is within shouting distance. Once the size of a force becomes large enough that means other than direct face-to-face contact are required, VHF combat radio is commonly used to provide a single voice channel for command and control. At the highest levels, where the required range of communications exceeds the line-of-sight range of VHF combat radio and higher capacity is required, trunk communications systems are used to provide long-range, high-capacity communications.

One major implication of dispersal of a formation over a very large area is the increased distances over which it is required to communicate. At brigade and possibly even battalion, these distances will exceed the range of ground-based VHF equipment, requiring the use of (possibly airborne) relays. Similarly, logistics units are strained by the longer times taken to drive between units deployed in the area of operations. Forces that are very fast moving can create similar difficulties, as was the case for the British division in the Gulf War. In support of one division, consisting of only two brigades,

Table 3.3

Characteristic Distances for Communications Within Units and Formations

	Conventional	Dispersed
Squad	200m	200m
Platoon	500m	500m
Company	1 km	1 km
Battalion	4 km	Up to 600 km
Brigade	12 km	Up to 1,000 km
Division	25 km	—
Corps	50 km	—

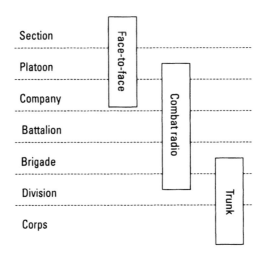

Figure 3.11 Typical army communications structure.

most of the communications resources of a corps were allocated, and these were stretched to the limit [12].

The requirement to be able to conduct different types of operations raises issues for both equipment and training. For modern armies, large-scale, high-intensity, combat operations tend to drive most equipment procurement. Most armies cannot afford to purchase different equipment for use in lower-intensity operations. Even if they could afford two (or more) sets of different equipment, these would likely require different training, suggesting that they would be operated by different specialist troops, a situation that is unlikely to be acceptable. As discussed in more detail in Chapter 5, it is generally required that the differences in equipment and specialist communications personnel used by a force for different types of operation are minimized.

3.7 Summary

Military organizations around the world are called upon to perform a large variety of tasks, ranging from the warlike to humanitarian. The ability to carry out this range of tasks is supported by the ability of armies to deploy to a remote location and provide their own support and logistics, including communications. The wide variety of tasks, and the associated need to operate in a traditional high-intensity, high-density environment, as well as operations in a widely dispersed mode pose special challenges for communications. An effective, modern architecture for military communications, therefore, needs to support command and control across the spectrum of operations, and the associated range of high- and low-density deployments. This architecture is the principal concern of the remainder of this book.

Endnotes

[1] "Joint Vision 2020," Director of Strategic Plans and Policy, J5 Strategic Division, Washington, D.C.: U.S. Government Printing Office, June 2000.

[2] Short, M. C., quoted in *Proc. AFA Air Warfare Symposium 2000*, Arlington, VA: Aerospace Education Foundation, Feb. 24–25, 2000.

[3] Ripley, T., "UAVs over Kosovo: Did the Earth Move?" *Defense Systems Daily*, Cirencester, U.K.: Defence Data Ltd., Dec. 1, 1999.

[4] Relevant U.S. doctrine is contained in U.S. Army Field Manual FM 7-8, "Infantry Rifle Platoon And Squad," April 1992.

[5] U.S. doctrine pertaining to the rifle company is contained in U.S. Army Field Manual FM 7-10, "The Infantry Rifle Company," Dec. 1990, amended Oct. 2000.

[6] See, for example, U.S. Army Field Manual FM 7-20, "The Infantry Battalion," April 1992.

[7] U.S. doctrine for the infantry brigade is found in U.S. Army Field Manual FM 7-30, "The Infantry Brigade," Oct. 1995, amended Oct. 31, 2000.

[8] Relevant U.S. doctrine can be found in U.S. Army Field Manual FM 71-100, "Division Operations," Aug. 1996.

[9] See, for example, U.S. Army Field Manual FM 11-50, "Combat Communications Within the Division (Heavy and Light)," April 1991.

[10] U.S. doctrine for mechanized troops is contained in:

U.S. Army Field Manual FM 7-7, "The Mechanized Infantry Platoon and Squad (APC)," March 1985.

U.S. Army Field Manual FM 7-7J, "Mechanized Infantry Platoon And Squad (Bradley)," May 1993.

[11] This rule was originally suggested by Miller, and has been empirically verified by many researchers since. See Miller, G. A., "The Magical Number Seven, Plus or Minus Two: Some Limits on Our Capacity for Processing Information," *The Psychological Review*, Vol. 63, No. 2, 1956, pp. 81–97.

[12] Rice, M., and A. Sammes, *Command and Control: Support Systems in the Gulf War*, London: Brassey's, 1994, p. 25.

4

Development of Tactical Communications

4.1 Introduction

Early commanders were constrained in their sphere of influence by how far they could move away from their troops and still remain in physical contact. This restriction on their ability to command and control placed a limit on the number of troops that could be placed under their command. A larger number of troops could be commanded through the development of a chain of command that provided a hierarchical structure within which orders could flow up and down between superior and subordinate commanders. Each commander still had to remain in physical contact with superior and subordinates, however. To embark on large-scale military endeavors, commanders needed to communicate effectively over distances larger than shouting distance and some form of communications system was required. Early systems were rudimentary, utilizing visual and acoustic signaling as well as messengers. The introduction of telegraphy and telephony saw the beginnings of a major change in command and control capability, and tactical communications systems have grown in complexity and scale to become an indispensable component of a commander's warfighting capability [1].

This chapter begins by briefly examining the early history of military communications and then focuses on the development of the two major tactical communications subsystems that are deployed by all major modern armies. From this background, a number of fundamental principles of battlefield communications are identified to provide a detailed understanding that can serve as a basis for subsequent analysis of the communications support that can be provided to modern tactical commanders.

4.2   Early History of Battlefield Communications

4.2.1   Early Military Communications

It made sense for early commanders to extend their controlling capability by using some extension of their physical means of communication. Early communications systems therefore took one of two simple forms: the *relaying of a message* by runner or courier, or *signaling* by some visual or acoustic means.

For several thousands of years messages between commanders have been carried by runners, either on foot or mounted on horseback. Messengers provided one of the major means of communications throughout both world wars and still provide an essential service today in transmitting bulky information around the battlefield. The means of transport has changed considerably, however. The horse was used until the end of World War I and was gradually replaced from the early 1900s by motorized transport in the form of the motorcycle and motorcar and, more recently, by aircraft.

Apart from human couriers, a number of animals have also often been used to transport messages [2]. For thousands of years, pigeons have been used to carry a message in a small tube attached to one leg. The British made the first large-scale use of pigeons during World War I with tens of thousands of birds in service. Although the birds were also used in World War II by infantry, armor, and aircrew, pigeons have not been used seriously since [3]. During World War I, the Germans and the British also briefly used dogs as messengers.

Early communications systems also extended ranges by relaying a message by shouting it between men stationed every hundred yards or so. Cannon shot, trumpet blasts, and drumbeats were also used to transmit simple messages over longer ranges. However, most early communications systems made much more use of visual signals than acoustic due to the longer ranges possible. Fire, smoke, rockets, flags, mirrors, and windmills were all used as simple signaling systems, and signaling range was often increased by building towers. Longer ranges were also possible after the invention of the telescope in the early seventeenth century.

In the late seventeenth century, a number of optical telegraph systems [4] were developed to transmit signals by placing mast-mounted beams or discs in different positions. For example, the French Radiated Telegraph machine could transmit 196 different signals (letters, code words, or phrases). Incidentally, communications were secure, since a codebook was required to decipher the signal. Optical telegraphs rapidly became obsolete in the early eighteenth century with the invention of electrical telegraphy, although some relatively sophisticated optical systems such as the heliograph, signaling lamp, and semaphore flags continued in service until the end of World War I [5]. After the invention of the electric telegraph, almost all other forms of communication were quickly replaced by systems that made use of electrical signals.

4.2.2 Electrical Telegraphy and the Telephone

The first major tactical and strategic use of the electrical telegraph was during the Crimean War (1853–1856). Strategically, a submarine cable laid from Varna to Balaclava assisted in connecting the British and French commanders with London and Paris, respectively. The cable provided great frustration for both commanders as, for the first time, both were in intimate contact with their political masters. Tactically, the Crimean War saw the first deployment of a telegraph troop with two telegraph wagons, a cable cart, a plough, and 24 miles of copper wire. By the end of the Crimean campaign, some 21 miles of cable had been laid, interconnecting eight headquarters.

In 1859, the Spanish and French armies made use of electrical telegraphy, albeit with civilian equipment and civilian operators. In 1860 the Italian Army made the first use of purpose-built military telegraph equipment and military operators. During the American Civil War, both the U.S. Army Signal Corps and the Confederate Signal Corps made use of electrical telegraphy.

In addition to telegraphy, staff officers began to demand telephones, which had become more common in civilian life. The first use of telephony by the U.S. Army was during the Geronimo campaign in Arizona in 1886. Although the telephone was also adopted by other armies in the late nineteenth century, it did not develop into the important tool it is today until the early twentieth century. The Japanese made the first extensive tactical and strategic use of the telephone during the Russo-Japanese war (1904–1905).

During World War I, line telegraphy provided the major means of communication. Most lines were buried well below the surface to protect them from artillery. The immobility of the buried cables and the sheer number of them forced better cable planning, and a grid system of main arteries into which units and formations could connect was developed. Each divisional area had a main artery with switching and testing centers connected to the main arteries of the division on either flank, thereby forming the grid system. Hundreds of miles of cables were deployed to support any advance until the grid system could be extended to cover the new positions. Tactical circuits were then connected to strategic telegraph circuits to allow the transfer of orders and information from headquarters to field commanders.

Telegraphy was the major means of communication during World War I, although telephony also grew in popularity since it gave staff officers timely, personal contact. So, in addition to line testing centers, telephone-switching centers sprung up all over the battlefield.

4.2.3   Radio

At the beginning of the twentieth century, line communications were augmented by the new technology of wireless communications. Although the British Army introduced a reasonably reliable set in 1915, radio telegraphy was not readily accepted throughout World War I, as the technology was still immature and radio telegraphy was sometimes inefficient, mostly due to a poor understanding of the physical processes involved and the low frequencies used. Despite that, radio telegraphy was much more flexible than line telegraphy and it was not long before tasks such as gun registration were being conducted by radio instead of line. For most tasks, however, radio was not well accepted in a static war where the telephone worked well.

At the beginning of World War I, all communications were via line from Army HQ down to company level. Radio communications were very limited. Communications followed the chain-of-command and had the same hierarchical structure. Communications from division to battalion levels were provided by the divisional signal company. Communications from battalion to company and below were provided by regimental signalers. By the end of the war equipment had improved, but not much had changed doctrinally except that the use of radio had increased to provide alternative means to line, particularly in communications below brigade where mobility was important. Line, however, remained the primary means of communication, with tens of thousands of miles of cable being laid by all sides. Line was still unreliable in some theaters and heavy use of dispatch riders or runners was made.

4.2.4   Between the Wars

Between the wars there were a number of developments in technology. The perfection of the vacuum tube allowed the consistent amplification required for AM. This allowed the first voice radio sets, or radiotelephone sets, to be introduced in 1918. Another important advance between the wars was the teletypewriter, or the printer telegraph. The teletypewriter required more power and was more complex to maintain than the Morse telegraph, but it was more accurate, faster, and relatively simple to use. The field telephone set was also developed between the wars and the Germans developed a small switchboard, which was rapidly copied by most armies.

During the 1930s, radio sets were developed to meet the needs of the infantry, artillery, armor, and aviation corps to provide the necessary mobility, range, and reliability. In 1934, the United States developed the 25-pound radio that became the first walkie-talkie. Between the wars radio sets became smaller and were carried on a number of platforms. The horse was finally replaced by the motorcar, leading to the mechanization of both radio wagons and cable wagons.

4.2.5 World War II

During World War II radio become ubiquitous across the battlefield, used extensively at the tactical level for the first time. The first armored command vehicle appeared and radios provided the necessary communications within highly mobile forces, often widely dispersed. FM radio was developed to provide noise-free communications. The infantryman was issued with a man-portable radio, which greatly enhanced operations. At the beginning of the war, only a few radios had been provided for catering to the main command links of formations, for divisional artilleries, and for internal use in armored and artillery units. By the end of the war, radio was used to conduct all essential tactical and administrative communications. The main reason for this was the inability of the line to keep up with highly mobile, widely dispersed forces often operating in inhospitable terrain. Headquarters required at least two operational command links, one for telegraph traffic and the other for voice. In addition to command, independent radio systems had developed for intelligence, air support, artillery, engineers, supply, and other services.

Line communications continued during World War II, although mostly for telephony work. World War II saw the integration of all forms of line and radio communication into high-quality links, regardless of the medium. Multichannel trunk radio made its first appearance in during the war, albeit in a simple form with up to eight duplex channels being provided by the British No. 10 set, one of the few World War II radios to operate in the UHF range. During the rapid advance eastwards from Normandy toward the end of the war, the line could not keep up and radio relay began to provide its utility on the support of fast-moving operations.

Although there were very few advances in communications technology in World War II, at the end of the war the face of military communications had changed considerably. While line was still utilized as an important medium, the mobility and dispersion of the battlefield had reversed the World War I situation so that radio had become the prime means of communication and line was only used as a secondary means when time allowed it to be laid. The communications provided by divisional signal units were higher-capacity radio and line links from division to brigade and brigade to battalion. Below battalion, lower-capacity radio and line links were provided by regimental signalers.

4.2.6 Current Doctrine

Two distinct battlefield communications systems have therefore developed to support the tactical commander. The first, above battalion, required high capacity links provided by the formation to interlink supported units with headquarters. The links were usually duplex and were limited to be from one unit to another. For example, a headquarters had a link to each of its subordinate units. These types of infrastructure links became known as *trunk communications* in line with their commercial equivalents. More commonly, as the links became more radio than line, they became known as *trunk radio*. Still, line is an important medium on the modern battlefield, and is laid within headquarters by hand reel and over larger distances by vehicle.

The second type of communications developed to allow units at battalion and below to perform tactical tasks. These links were flexible and responsive, under the direct control of the commander. Links were established using single-frequency, half-duplex, all-informed radio nets allowing the commander maximum flexibility to command a number of subunits. These types of communications have become known as *single channel radio*, or more commonly as *combat net radio*.

Overlying both of these services was still the requirement to send and receive bulky communications. These needs were still being met by dispatch riders of the signal dispatch service (SDS) or postal service. As we saw earlier, SDS is provided by a wide variety of forms of military transport: motorcycle, vehicle, and aircraft.

So, two broad types of communications service have evolved: combat net radio, or single channel radio, and trunk communications (including line, radio, and SDS). Doctrinally these divisions still exist on the modern battlefield, albeit in more sophisticated forms.

- *CNR subsystem.* The CNR subsystem is a ruggedized, portable radio (HF, VHF, and UHF) network carried as an organic communications system for combat troops (brigade level and below). Radios are invariably interconnected to form single-frequency, half-duplex, all-informed, hierarchical nets, providing tactical commanders with effective support to command and control.

- *Trunk communications subsystem.* This subsystem provides high-capacity communications links down to brigade level. The subsystem traditionally comprises multichannel radio equipment, line, switches, and terminating facilities to provide voice, telegraph, facsimile, and data communications, as well as a messenger service.

These two subsystems are described in more detail in the following sections. It should be noted that in U.S. doctrine, a third subsystem, the Advanced Data Distribution System (ADDS), is provided—the need for this third element is discussed in Chapter 5, and a more detailed description is provided in Chapter 8.

4.3   CNR

All CNR stations on a particular net operate on the same frequency in a shared channel, as illustrated in Figure 4.1. Each station can transmit and receive, but not at the same time. CNR systems are therefore often referred to as single-frequency, half-duplex in nature. To access the net, all stations operate using a simple CSMA-like procedure. When the net is free, any station can initiate a call. However, since the communications channel is shared in a half-duplex manner, users must use a protocol to determine the right to talk at any particular time. In traditional voice communications, this protocol [called *voice procedure* or *radio telephone (RATEL) procedure,* or *radio operating procedure*] [6], is based on the use of call signs for the beginning of conversations and keywords for handing over the right to speak, and for the termination of conversation. Additionally, a higher level of control is normally forced on the network by one of the stations called the *net control station* (NCS), which is responsible for running the net, including net discipline and control of frequency changes.

The significant military advantage of such a net is that it is *all-informed* in that each station receives all transmissions from all other stations whether they are intended for it or not. This configuration is essential for command

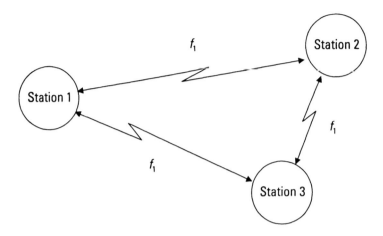

Figure 4.1 Single-frequency, half-duplex net.

and control because it allows commanders to pass orders efficiently to a number of subunits without having to repeat calls. Similarly, commanders can monitor activities by monitoring the net traffic without specifically having to request a report. The all-informed nature of the CNR net does have a limitation, however, due to the requirement for all stations to be able to hear the transmissions from all other stations. The range that the net can cover is therefore limited, although stations can be used to pass on transmissions if one of the stations can only hear a limited number of transmissions. Extended ranges can also be obtained by the use of manual or automatic rebroadcast stations, which can provide coverage into other areas.

Single-frequency, half-duplex operation is the most suitable for CNR since it provides the maximum flexibility and survivability. To provide these qualities, however, in addition to the other military requirements of reliability, security, and capacity, individual military radios designed for CNR systems are generally expensive, heavy, and bulky and will normally constitute almost a full load for one individual.

CNR is normally provided in three frequency ranges: HF, VHF, and UHF. HF CNR is normally provided with a frequency range of 2 to 30 MHz. Voice is transmitted in a 3-KHz channel using SSB. VHF CNR in most armies is capable of operating in the frequency rage of 30 to 88 MHz, although greater ranges are sometimes encountered. Channel spacing in older analog systems is 50 kHz, but is now commonly 25 kHz providing FM voice. Data is transmitted using a form of FSK, to provide a signaling rate of 16 Kbps in the available 25-kHz channel.

Although CNR nets are single-frequency, all-informed nets, they are operated in a hierarchical manner and are generally used to reinforce the chain-of-command. That is, the NCS is normally collocated with the commander and the net is normally viewed as having the architecture shown in Figure 4.2.

When nets are employed to support a military command structure, commanders are invariably on at least two nets to remain in contact with their superior as well as their subordinates. Figure 4.3 illustrates a simplified version of the command nets for a division. Note that each commander requires access to two radios, which are not necessarily connected. The implications of this hierarchical net structure are discussed further in Chapter 5.

Use of the RF spectrum on the battlefield is dictated by a trade-off between capacity, mobility, and range. These issues have a considerable impact on the use of radios on the battlefield and are discussed in much more detail in Chapter 5 [7]. For the moment, Table 4.1 lists the traditional applications of the RF bands to use in CNR nets.

As discussed in Section 4.2, radio was first used on the battlefield for trunk communications as an alternative to line. Soon, however, radio was used for previously difficult tasks such as connecting forward observers to artillery batteries, which avoided the laying of hundreds of miles of cable to support major offensives. As sets and antennas reduced in size, they began to be employed to form artillery-infantry nets and infantry-armor nets. By the end of World War II, the United States had deployed CNR to most elements of the infantry and other arms.

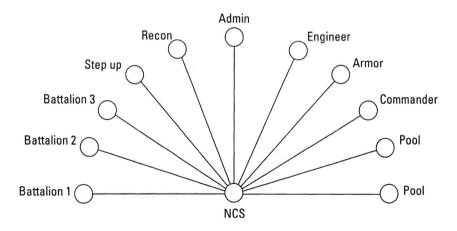

Figure 4.2 CNR net diagram for a notional brigade command net.

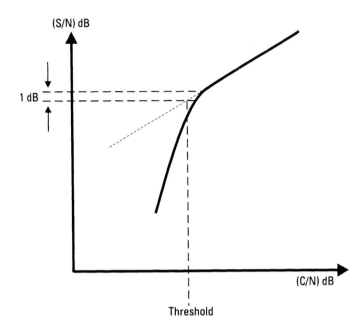

Figure 4.3   CNR net hierarchy for a notional division's command nets.

Since CNR was first deployed, there have been a number of advances in technology. Electronic design has improved dramatically and modern CNR makes use of frequency synthesizers, integrated ATU, and more efficient antennas. Synthesizer design, in particular, has allowed radios to be capable of tuning across a number of bands. Sets have become lighter, smaller, and more reliable due to smaller integrated-circuit components, which require less space, power, and maintenance. There have been some improvements in battery technology that reduces the overall weight of the radio, although the battery remains the heaviest component of the system. Cryptographic equipment has become smaller, and is now commonly integrated into the set or handset. Over-the-air rekeying has eased considerably the burden associated with cryptographic key management and distribution across the battlefield. Finally, better modems have allowed CNR to operate at higher data rates, although the radio is still fundamentally a terminal on a voice network and is therefore not ideally suited to acting as a data network node.

Despite these evolutionary developments, the doctrinal use of CNR has remained largely unchanged since World War II. The major difference is in the ability to pass data over most CNRs, although most in-service CNRs are still analog radios and are not well placed to cope with the expansion in the

Table 4.1
Application of RF Bands to CNR Nets

Band	Application to CNR Net	Advantages	Disadvantages
HF	Company, battalion, brigade, command, and administration nets Rear links Strategic interfaces Armored, engineer, artillery, and reconnaissance nets Special forces patrols	Cheap, man-portable, long-range Graceful degradation of communications	Short surface-wave range from man-pack and vehicle whips Relative immobility for sky-wave communications; Limited number of frequencies and small bandwidth available
VHF	Platoon, company, battalion, brigade, command, and administration nets Armored, engineer, artillery, reconnaissance nets	Represents optimum trade-off between bandwidth, power, weight, and size Sufficient bandwidth available for encryption Higher quality	Weight, particularly including encryption Range limited to radio horizon because of low antennas
UHF	Squad and platoon nets Artillery between guns Ground-to-air nets Special forces patrols	Lightweight, hand-held Short-range, which reduces probability of intercept	Line-of-sight limits range Not easily interfaced to VHF CNR

From: [8]

volume of data expected to support concepts such as network-centric warfare. These issues associated with CNR on the digitized battlefield are discussed in more detail in Chapter 7.

4.4 Trunk Communications

Within a major headquarters or logistics installation, commanders and staff officers are connected by means of local links to a central headquarters hub. The links may comprise a LAN (for data), local loops (for telephony) or perhaps a single converged system. The hubs provide subscribers with connections to other subscribers within the headquarters using the local network; as well as with remote access to the combat radio network and access to the trunk network. The following sections provide a brief description of the development of trunk networks from rudimentary chain-of-command networks providing voice and telegraph, to modern area trunk systems providing the full range of subscriber facilities.

As illustrated in Figure 4.4, first-generation trunk networks provided communications links that followed the chain-of-command. This arrangement made sense, as it provided communications links that supported the flow of information, which was between commanders. Commanders also saw this as a natural arrangement as the doctrinal flow of information was up and down the chain-of-command. In fact, doctrine generally prohibited—and in most armies still prohibits—any communications outside the chain of command.

Direct chain-of-command networks had the serious disadvantage, however, that each headquarters was required to act as a tactical base as well as a communications node. These two roles are mostly in conflict. A communications node requires access to sufficiently high terrain to provide the range necessary to reach the superior headquarters as well as to the headquarters of subordinate units. In a tactical environment, in contrast, the commander must

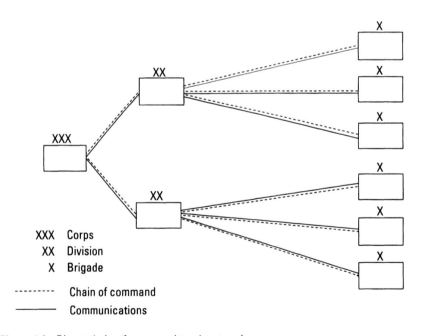

XXX   Corps
 XX   Division
  X   Brigade

-------- Chain of command
———— Communications

Figure 4.4   Direct chain-of-command trunk network.

conceal the location of the headquarters since it is too vital an asset to have perched on a hilltop. The collocation of the communications equipment with the headquarters also constrained the mobility of the headquarters and increased its vulnerability to both visual and electronic detection. Additionally, any movement of the headquarters or damage to the network caused a disproportionate disruption to communications. For example, if the communications are lost to brigade headquarters, there is no mechanism for communications between the divisional headquarters and that brigade's battalions.

Figure 4.5 provides an example of a *second-generation trunk network*, which alleviated earlier difficulties by taking the logical step of displacing communications from the chain of command. By creating a physically separate communications site (often known in the British system as a communications center) that communicated to the headquarters over short cable or radio links, a separation of the tactical and communications roles of the headquarters was achieved. As a result of this separation, tactical headquarters and communications sites could be planned with a higher degree of independence, although the headquarters was still constrained by having to be near its communications center.

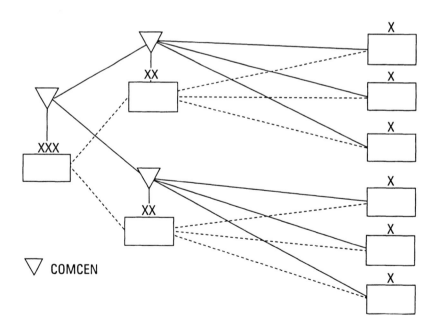

Figure 4.5   Displaced chain-of-command trunk network.

Examples of this type of network include the early networks deployed after World War II, particularly the British Army's BRUIN network deployed in northwest Europe from the 1960s until 1985 when it was replaced by PTARMIGAN.

Although the siting difficulties were alleviated in the displaced chain-of-command network, any movement of the headquarters (or communications center) or damage to the network still caused a disproportionate disruption to communications. The next logical step was to improve the reliability of the network of the headquarters, providing a second communications center, which provided a second communications link into each headquarters and allowed one of the communications centers to be destroyed or moved without disrupting communications. The network could then be reconfigured without disrupting communications, thereby improving reliability. Figure 4.6 illustrates an example of this third-generation, *expanded chain-of-command network.*

The second communications center also allowed the headquarters to move using a process called *step-up* in which one-half of the headquarters deployed to the new location with one of the communications centers, set up, and established communications with the old headquarter location and the superior and subordinate headquarters. Command was then transferred to the new headquarters location and the old location was packed up with its

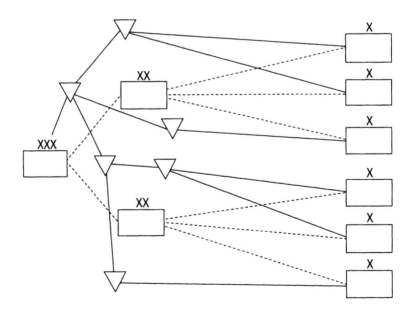

Figure 4.6   A simple expanded chain-of-command trunk network.

associated communications center and moved to the new headquarters location. Perhaps the best example of a third-generation network is the U.S. Army's Army Tactical Communications System (ATACS), which was deployed until the mid-1990s when it was replaced by the Mobile Subscriber Equipment (MSE) trunk system.

4.4.1 Generic Fourth-Generation Trunk Communications Architecture

The logical extension of these developments is the fourth-generation area trunk network illustrated in Figure 4.7. Most modern trunk communications systems have been developed as fourth-generation networks. An area trunk network provides a grid (or mesh) of switching centers deployed to provide coverage of the area of operations. Nodes are interconnected by bearers, which are traditionally multichannel radio-relay links in the UHF or SHF band. Headquarters connect to the nearest trunk node by radio relay and can then have access to any other headquarters that is also connected to the network.

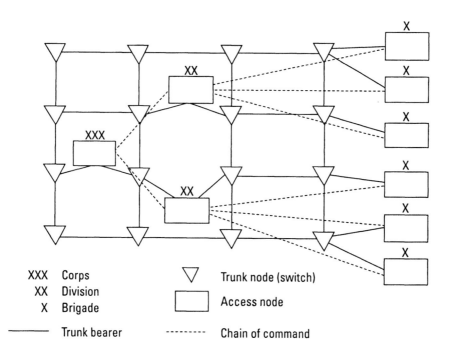

Figure 4.7 A simple meshed area trunk network.

An area network contains significant redundancy and can sustain considerable damage, or cope with substantial movement, because of the alternative routes available. A key advantage of this network topology is its capability to automatically cope with traffic routing changes resulting from movement of subscribers or network outages. Nodes can be moved rapidly to reconfigure the network as required by the tactical situation. Commanders are no longer constrained by their communications links and can deploy as required by the tactical situation with the only siting constraint that the headquarters location must be able to communicate to at least one of the trunk nodes.

The obvious disadvantage of an area network is the significant amount of equipment and manpower required to establish the large number of redundant nodes. A first-generation network for a deployed corps required approximately 12 nodes for the major units; a fourth-generation corps network typically deploys 40 or 50 nodes. This disadvantage is by far outweighed, however, by the greatly improved flexibility, reliability, survivability, and capacity provided by an area trunk system.

4.4.2 Components of Trunk Networks

While each nation implements their trunk networks in slightly different ways, there are many common elements. The *trunk node* is the basic building block of the trunk network architecture. Nodes are connected together by multichannel *radio relay* links to provide a meshed infrastructure of trunk nodes. Headquarters and command posts connect to the network, and hence to each other, through *access nodes. Single channel radio access* (SCRA) provides full network access to mobile subscribers, and combat-net radio users are able to have a relatively limited interface to the network through a *CNR interface* (CNRI). Other trunk networks are accessed through a *tactical interface installation* (TII). Each of these components is discussed in more detail in the following sections. Components are discussed in generic terms, not related to any particular one of the national trunk networks listed in Table 4.2 [9].

4.4.2.1 Trunk Nodes

As described earlier, the trunk node is the basic building block of the trunk network. Nodes are deployed and maneuvered to provide an area trunk network that allows combat units to deploy and maneuver as the tactical situation allows. Figure 4.8 illustrates the deployment of a trunk network in a generic deployment. Note that, as required by convention, the network

Table 4.2
National Trunk Networks

Country	Trunk Network	Designation
Australia	PARAKEET	PARAKEET
France/Belgium	Réseau Intégré de Transmissions Automatique	RITA
Germany	Automatisierte Korps-Stammnetz	AUTOKO
Italy	SOTRIN	SOTRIN
The Netherlands	Zone Digital Automatic Communications	ZODIAC
United Kingdom	PTARMIGAN	PTARMIGAN
United States	Mobile Subscriber Equipment	MSE

provides an interface to the network on the right, and has interfaces provided by the higher formation and the formation on the left flank.

Of course, Figure 4.8 does not accurately reflect the number of trunk nodes likely to be deployed in each area. Typically there are approximately 40 trunk nodes in a corps network, with approximately four allocated to each division and the remainder allocated as corps assets. Figure 4.8 also shows a physical grid pattern of deployment. While the network is invariably deployed with a logical-grid connectivity, the location of the trunk nodes is dictated by the terrain and the tactical situation.

Nodes are deployed and redeployed by the network managers to adjust to the needs of combat forces as dictated by the tactical battle. Nodes can be redeployed to facilitate an advance or withdrawal, or movement to a flank. The density of the network can also be modified to provide sufficient capacity to cope with any changes in force composition and location. Figure 4.9 illustrates the basic components of a trunk node:

- *Switch.* The nucleus of each node is the switch vehicle, which contains a processor-controlled digital switch. In most modern networks the switch is currently an automatic circuit switch that incorporates an embedded packet switch. Current switches typically provide 16 or 32

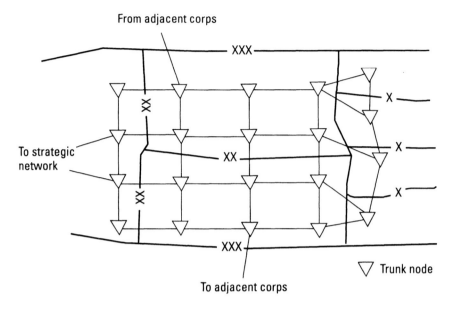

Figure 4.8   Generic deployment of trunk nodes in a corps area.

internodal trunk channels. One channel is generally allocated to engineering and the remainder are bulk-encrypted and as internodal trunks. The meshed area network is created by connecting each node to at least three others through radio relay bearers.

- *Node operations center (NOC)*. The NOC contains an operator interface to assist in engineering the switch and trunk encryption equipment as well as to allow some limited patching between trunk channels. The NOC would normally be located in the switch vehicle.

- *Network management facility (NMF)*. The NMF performs link management for the radio relay links connected to the switch, including engineering of the links and frequency management. Normally some functions are also performed on behalf of the next level of management in the network. The NMF is normally a separate vehicle manned by the trunk node commander.

- *Radio-relay detachments*. In most modern networks, four or five radio relay detachments are typically deployed with each trunk node. Each detachment can terminate radio links from approximately three other radio relay detachments at either another trunk node, an SCRA *radio*

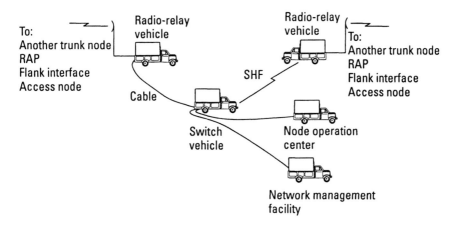

To:
Another trunk node
RAP
Flank interface
Access node

Radio-relay vehicle

Radio-relay vehicle

To:
Another trunk node
RAP
Flank interface
Access node

SHF

Cable

Switch vehicle

Node operation center

Network management facility

Figure 4.9 The basic components of a typical trunk node site.

access point (RAP), a flank interface, or an access node. Detachments are normally sited on a suitable nearby hill and are connected to the node switch by a cable "tail" or, when the terrain does not allow cable to be laid, by SHF "down-the-hill" radio. Each node switch is normally connected through radio relay to other node switches to ensure a robust, survivable network, as illustrated in Figure 4.7.

4.4.2.2 Bearers

Any wideband communications medium could potentially provide the multichannel, digital links between trunk nodes. Although sufficient bandwidth is available from coaxial cables, waveguides, and optical fibers, line of any type is very time-consuming to deploy and recover. Radio systems are far more flexible and are therefore normally the preferred method of linking trunk nodes.

4.4.2.2.1 Radio Relay

Frequency of operation. The most favored radio-relay systems operate in the VHF/UHF and SHF bands in order to provide the necessary bandwidth to interconnect trunk nodes. Within those bands, the frequency of operation of radio relay for a particular trunk network is selected as a trade-off between the need for high capacity (and therefore high frequency) and ease of antenna alignment (and therefore low frequency). At the high-frequency end, SHF systems are constrained by line-of-sight and limited to less than 12 GHz due to strong absorption by rain and other precipitation. For radio-relay applications, SHF systems therefore require careful selection of sites, high masts, and highly directional antennas, which require a long time to erect and align. These requirements normally prohibit SHF systems from meeting tactical mobility constraints that demand a short time into and out of action. SHF does have application, however, for short-range applications in down-the-hill shots from radio-relay vehicles to trunk nodes. For radio-relay applications, the VHF/UHF band is favored as a reasonable military compromise between adequate channel capacity and tactical mobility. Three bands are commonly utilized: Band I (225–400 MHz); Band II (610–960 MHz); and Band III (1,350–1,850 MHz).

Antennas.   Radio-relay links are point-to-point, and directional antennas are used to make the best use of available power and maximize gain in the intended direction. Directional antennas are also very useful because they reduce the radiation in unwanted directions, which in turn minimizes the possibility of interception as well as reduces interference on radio systems in the near vicinity. For the receiver, a directional antenna minimizes the interference from other transmitters, including hostile jammers. The gain of these antennas is limited to about 10 to 25 dB so that the alignment and concealment are straightforward and the antenna is robust and light enough to erect quickly.

Siting.   Radio-relay terminals located on high ground will give the best possible radio path. For tactical reasons, headquarters and trunk nodes are usually sited in more concealed positions. Since the two sites may be some distance apart, the radio-relay site is normally connected to the asset it is serving by a cable or, for longer distances, an SHF down-the-hill radio link. Sites are normally selected from a map reconnaissance and confirmed by a path profile analysis. This produces a short list of possible sites that are then subjected to a physical reconnaissance considering such factors as local obstacles not shown on the map; possibilities for concealment and camouflage; access for vehicles; and whether the site has already been occupied, since high features are always attractive for many other force elements, including radar, surveillance, EW, and communications detachments (noting that they are also attractive to adversary artillery units).

Radio-relay range extension.   As noted earlier, although UHF and SHF frequencies provide sufficient bandwidth, they have the planning difficulty of requiring line-of-sight between antennas. Additionally, to maintain the required signal-to-noise ratios, particularly for data links, radio paths are limited to planning ranges on the order of 20 to 30 km. Often, therefore, two trunk nodes cannot be connected by one link and, as illustrated in Figure 4.10, an intermediate relay station, called a *radio-relay relay* station, is inserted in the link to extend range. This relay is sited so that there is a good line-of-sight path to both radio-relay terminals so that signals received from each terminal can be automatically retransmitted to the other. Where long internodal links are required to support widely dispersed forces, there may be several relay stations, although satellite or troposcatter links are generally preferred on long paths.

4.4.2.2.2 Tropospheric Scatter

From an application point of view, the most interesting characteristic of troposcatter radio links is the great distance over which reliable communications can be obtained without the need for intermediate repeaters. For example, troposcatter ranges would be on the order of 150 to 200 km, as opposed to tens of kilometers for radio relay. This feature is particularly useful for cases in which the terrain is difficult such as in connection between remote sites such as those in the desert or jungle; connection of a remote island to the mainland or another island, or across uncontrolled territory or geographic obstacles; and networks in which the possibility of sabotage at unattended repeater stations should be avoided.

The possibility of avoiding repeater stations, with their equipment, antennas, buildings, roads, and problems of accessibility and maintenance, may lead to a more economical solution with a troposcatter link instead of a traditional line-of-sight link. Troposcatter radio systems have particular application in less

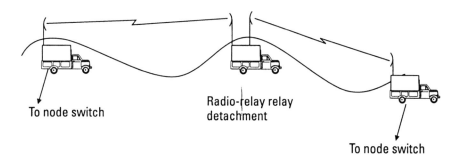

Figure 4.10 Range extension using a radio-relay relay station.

conventional military operations where the security of repeater stations is an issue.

Advantages of troposcatter systems.   The main advantages of troposcatter systems can be summarized as follows:

- They permit a long path. A single hop can have a length of hundreds of kilometers (five or six times the usual length of a line-of-sight link). They therefore fit neatly between line-of-sight communications systems and satellite communications systems, providing much more bandwidth than HF systems.

- They can be used on difficult terrain, which has little influence on the system design. This is particularly useful when the terrain has not been secured or, for example, when repeaters would have to be located above the snow line.

- Coverage of very large areas can be obtained with a small number of hops.

- Only a few repeaters are required because of the long hop-length.

- Fewer frequencies are required because of the reduction in the number of stations.

- Security against sabotage or catastrophic failure is easier to achieve, as there are fewer stations to protect.

- Procurement and operating costs are reduced because of the lower number of repeaters.

- There is a high immunity to interception because of very narrowbeam antennas.

Disadvantages of troposcatter systems.   The main disadvantages are as follows:

- The cost per station is high. However, this factor may be superseded by the advantages given above and the troposcatter solution may be the most cost-effective choice as the overall costs (investment, operation, and maintenance during the life of the link) may be less than any alternative solution.

- High-gain antennas are difficult to orientate accurately.

- Antennas must have an unobstructed view of the horizon.

- There is a high RF hazard from the high power output at the antennas.

• There is a risk of interference over a wide area if the same frequencies are used at other stations.

Troposcatter was used in a combat environment for the first time by U.S. forces in 1962 to provide a backbone communications system that extended the length of South Vietnam [10]. Interestingly, the only U.S. National Guard communications unit deployed to the Gulf War was equipped with the TRC-170 troposcatter equipment [11]. Troposcatter remains a useful bearer system as part of a trunk communications system.

4.4.2.2.3 Satellite Network Links

Fourth-generation trunk networks were developed to provide reliable, survivable communications across a corps area within a high-intensity deployment. With the demise of the Cold War, a broader spectrum of operations has placed greater demands on meshed-network architectures. The first major change came in the Gulf War, where despite supporting corps deployments, trunk networks had great difficulty in providing a meshed architecture while keeping up with the rapid rate of advance. In fact, networks were invariably deployed in a linear fashion, requiring switch software to be reengineered to cope with the increased length of the networks from one end to another. Radio-relay bearers are ideal for a high-density deployment, but do not provide a good bearer system for rapid movement of supported forces over large distances.

As U.S. and U.K. forces were investigating solutions to this issue, a similar problem had arisen for slightly different reasons in the various deployments in the former Republic of Yugoslavia. The provision of an area network had proven very difficult due to the inability to find secure sites for relays between widely dispersed units and formations.

The solution to both of these problems has been the introduction of satellite trunk links to extend the range of connections between nodal, which does not affect the logical layout of the area network, but dramatically increases the range of internodal links. As illustrated in Figure 4.11, area networks are provided locally where possible and satellite links are provided to link these network "enclaves" or "islands" together.

4.4.2.2.4 Commercial Communications Networks

In addition to satellite links, connections between nodes can also be extended by connections through commercial communications networks. However, commercial systems are not always available or survivable enough

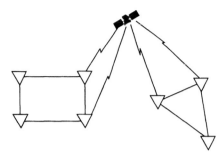

Figure 4.11 A simple area network architecture employing satellite internodal links.

for inclusion in trunk networks. Satellite trunk links are therefore preferred to span large distances between nodes.

4.4.2.3 Access Nodes

An *access node* is a processor-controlled digital switch capable of handling the number of fixed subscribers at the access point. The access node can switch calls between these subscribers or switch them to an outgoing trunk line through the trunk switch. Normally, two levels of access nodes are provided: a small access node for brigade headquarters and a large access node for divisional headquarters. In some networks, access nodes are provided down to lower levels to serve subscribers at regimental or battalion command posts. In most networks, however, access to these levels is provided through SCRA, because the users are mobile and fixed subscriber access to the network tends to be inappropriate. As illustrated in Figure 4.12, subscribers are connected to the access node, which is connected to the network (to a trunk node) through radio-relay bearers. Large access nodes (for divisional headquarters and above) are normally connected to two trunk nodes; small access nodes (for brigade headquarters and below) are normally only connected to one trunk node, with a standby link engineered to a second trunk node in case of failure of the first.

Small access nodes normally serve approximately 25 subscribers and the large access nodes provide for approximately 150 subscribers. As well as the ability to interface to radio-relay bearers, both types of nodes have the facilities to interconnect to satellite or troposcatter links, or to commercial carriers. Figure 4.13 shows a simplified layout of a small access node.

As illustrated in Figure 4.14, the large access node has a similar layout as the small node. In addition to the ability to cope with a larger number of subscribers, the large access node is connected to two trunk nodes. The large

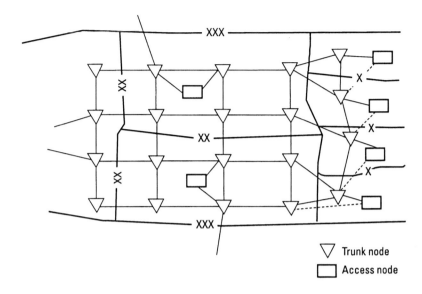

Figure 4.12 Connection of access nodes to the trunk network.

access node also contains an NOC and an NMF, similar to those found in the node center of a trunk node.

4.4.2.4 SCRA

Access nodes provide access for fixed network subscribers. Access for mobile or isolated subscribers is provided through the duplex VHF-radio access of the SCRA subsystem. The SCRA subsystem comprises *subscriber terminals* that provide mobile users with network facilities (e.g., voice, data, telegraph, and facsimile) equivalent to those available to static subscribers of an access node, and RAP through which subscribers access the trunk network. As illustrated in Figure 4.15, an RAP is normally connected to the network by connecting to a trunk node via cable, SHF down-the-hill radio, or UHF radio-relay link. A standby link is engineered to another trunk node to be used if the primary link fails. RAPs are deployed like PCS base stations to provide overlapping areas of coverage to provide continuous coverage for a mobile user.

RAPs. In most in-service networks, each RAP can accommodate approximately 50 affiliated mobile subscribers (i.e., 50 subscribers who are within the RAP's operating area). Not all affiliated subscribers can communicate at once, however, and an RAP can normally only cope with simultaneous calls from a dozen subscribers. While the RAP will have a planning range of 15 km,

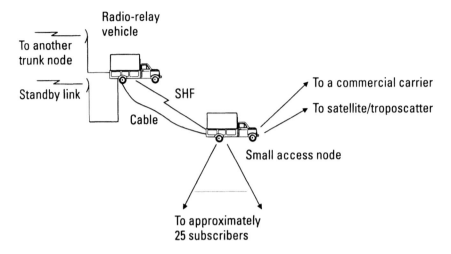

Figure 4.13 Simplified layout of a small access node.

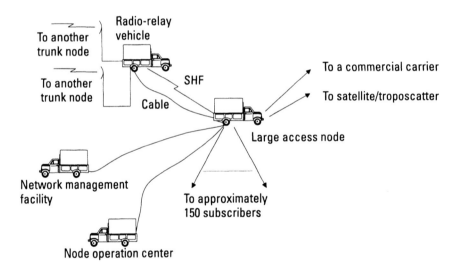

Figure 4.14 Simplified layout of a large access node.

the actual range between the RAP and subscriber will depend on the terrain and the heights of the RAP and mobile antenna. Figure 4.16 shows a simplified layout of an RAP site, which comprises an RAP vehicle containing transmitters and receivers connected to an omnidirectional antenna. RAPs

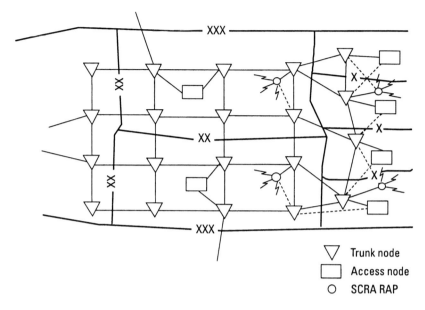

Figure 4.15 Connection of SCRA RAPs to the trunk network.

are normally able to operate on a range of power settings so that the lowest power level can be chosen for each user to reduce the electronic signature of the RAP.

Subscriber terminals. To connect to the network, a mobile subscriber is required to go through a process of *affiliation*, during which the subscriber's identity is validated. Each network conducts its affiliation process slightly differently, but normally the subscriber is required to perform some deliberate action to affiliate for the first time. In some networks the subscriber must continue to affiliate each time connection is required to a new RAP. If the subscriber does not know the number of the closest RAP, a search can be initiated to find the most suitable RAP. In more modern networks, reaffiliation is automatic as the subscriber is handed over from one RAP to another while moving through the network. Upon affiliation, the power output of the terminal is normally reduced to the minimum possible power level to ensure similar power levels at the RAP from all subscribers, and to reduce adjacent channel interference

Direct access. In the more modern networks, mobile subscribers in close proximity within the same RAP area do not have to communicate through

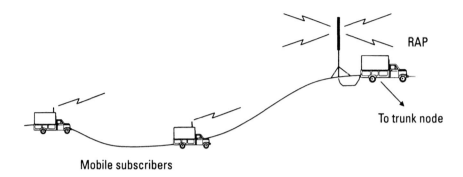

Figure 4.16 Simplified layout of an RAP.

the RAP and can contact each other directly. This not only reduces traffic load on the RAP, which is only used to relay the call if the direct connection cannot be made within a specified time, but it allows communications when the network is not present.

4.4.2.5 CNRI

SCRA is an extension of the trunk network in that subscribers have access to the same network facilities as static subscribers in command posts. In addition to SCRA, most networks provide an additional form of access to lower-level commanders who are not network subscribers but who can use their CNRs to have temporary access to the network. This CNRI provides a semi-automatic, voice interface between VHF and HF CNR users and the network. The range of CNRI depends on the CNR frequency band. In each frequency band, the CNRI vehicle has a hailing channel through which CNR users can contact the operator who sets up the call through one of the VHF or HF traffic radios. In the other direction, calls made from the network are automatic and do not require operator intervention. While it is possible for a CNR user to access the network in this way, CNR nets are single-frequency, half-duplex so that CNRI only provides rudimentary access to network facilities. RATEL and CNR net procedures must be observed by CNRI users, even those who are trunk subscribers.

4.4.2.6 Tactical Interface Installations

For NATO in Western Europe during the Cold War, it was important that each corps was able to provide communications to flanking formations from other nations. While the Cold War deployments are less likely, multinational alliances are essential in almost all modern deployments and it is perhaps even more important for modern networks to be able to interface to those of other nations. There are therefore a large number of standards and agreements that cover the interface of trunk networks. For example, within NATO, the internetwork interface for analog signals, voice, and telegraph is arranged through the standard agreement STANAG 5040 [12], and interoperability between digital EUROCOM [13] systems using 16-Kbps modulation is provided by STANAG 4208 [14]. It is unlikely, however, that national procurement programs will ever be completely synchronized and most networks arrange to support the interface through a specific installation called the TII.

4.4.3 Trunk Subscriber Facilities

Traditionally, trunk networks provide static and mobile subscribers with secure voice, data, telegraph, and facsimile services. In the future, these services might also include video. The following gives a brief overview of the equipment and facilities provided to subscribers in most current networks.

Subscriber equipment. Subscribers make voice calls through the digital circuit-switched network using a digital or analog telephone. Data, facsimile, and telegraph facilities are provided through the attachment of a *data adaptor.* The subscriber equipment performs analog-to-digital conversion to provide an output data rate, normally at 16 Kbps using a form of delta modulation.

Security. Trunk network security is provided through bulk encryption of the trunk links so that the network is normally treated as a SECRET-high network. That is, it is assumed that any subscriber at any point in the network is cleared at least to the SECRET level, and that there are physical security measures in place to ensure that only appropriately cleared personnel can receive the information. If higher classifications are desired, links between users can be individually encrypted once a circuit has been established. This form of double encryption is commonly used for intelligence and electronic warfare users.

Voice services. Subscribers can move about within the network and may not necessarily remain near the same subscriber equipment all the time. Most modern networks therefore give each subscriber a unique number called a *directory number* that they can enter on any phone in the network to tell the network where they can be found so that a call can be placed to them. A subscriber can be located at any time through *affiliation* and *flood search*. It does not make sense to have to issue a phone directory for an organization whose staff is changing constantly, so most networks also use the NATO 7-digit deducible directory (STANAG 5046 [15]), in which the subscriber's directory number is deducible from the parent unit and appointment. Trunk networks provide the normal network voice facilities such as call waiting, transfer, and call forward, along with the more modern facilities of compressed dialing, abbreviated dialing, and conference facilities. Particular military requirements are met by the *precedence* and *preemption* facilities. Preemption allows selected subscribers to manually override other callers for access to outgoing trunks (or if the called extension is busy), providing the caller's precedence is greater than that of the subscribers involved in the busy conversation.

Affiliation.   Subscribers are free to access the network from any point where there is a subscriber terminal. To do so, however, the network must know where each subscriber is located so that incoming calls can be routed to the correct location. Each subscriber is therefore required to affiliate to the network by entering an affiliation sequence into the nearby terminal, which tells the local switch where the subscriber is currently located.

Flood search.   When a call is placed, the subscriber is located by the network through a flood-search mechanism in which the switch to which the calling subscriber is connected calls all other switches until the called subscriber's parent switch responds. Once the subscriber's location has been identified, the connection can be made.

Data services.   Data services can be provided by a number of methods, but are normally provided by a packet-switched network integrated into the circuit-switched network. Chapter 6 provides more detailed discussion on the provision of voice and data services across modern networks. In current networks, however, there are two main methods available to integrate circuit and packet switching within the same trunk network:

- *Common trunking.* The most obvious approach is through common trunking, where the circuit and packet switches share the common transmission facilities via multiplexing equipment. However, integration at this level does not really provide much for improvement in transmission efficiency and utilization, or the exchange of traffic between different communities or terminal types.

- *Embedded network.* In most modern networks, the packet-switched network is embedded in the circuit-switched network. Each switch still has its own community of users, but the packet switches act as user terminals within the circuit-switched network. Using the connections through the circuit-switched network as bearers, each packet switch connects to others to form the packet-switched network. The allocation of trunk channels to the packet-switched network is normally predetermined, although many networks allow the allocation of additional capacity, should the amount of traffic between any pair of packet switches begin to increase.

4.4.4 Network Management

As trunk networks have matured and have become more independent of the chain of command, network management has also had to evolve as a separate function. As networks have become more complex, the management task has become an indispensable component of modern networks.

4.4.4.1 The Management Task

The principal role of network management is to ensure that the network provides and maintains effective and reliable communications. A number of management tasks must be conducted: assessment of the requirement, including identification of subscriber communities for voice, data, SCRA, and CNRI; formulation of a plan for network connectivity, including the radio path planning for bearers; planning for the location of key network elements such as radio-relay detachments and RAPs; allocation of resources to meet the plan; preparation and issue of orders; assignment of frequencies; management of changes to the network, both planned and unexpected; system performance analysis; traffic engineering; cryptographic management; system maintenance; and implementation of EP plans.

The impact of the role of network management cannot be understated. Although modern area networks are inherently flexible, they can only remain so if they are well managed. The network must redeploy, and key network elements will be moving, or may even be destroyed. The continual configuration and reconfiguration of the network must be tightly controlled if the network is to remain available to users. The trunk communications system is a critical element of the commander's combat capability and must be able to withstand the demands of the tactical environment. While the meshed area network provides great flexibility, it also introduces a significant overhead in terms of network management.

4.4.4.2 Functional Levels of Management

The evolution of tactical communications systems has seen a disassociation of communications elements from the units and formations they support. Similarly, it is not common for network management elements to have a close affiliation with formations. In a modern area network, the management structure is centralized to ensure the optimum usage of system resources and the provision of maximum availability to users. That does not mean to say, however, that there is not a hierarchy of management. Rather, there are a number of functional levels of management and management elements are normally associated with logical, geographical groupings of network assets to provide the redundancy necessary for a survivable management system. The number and the nature of functional levels of management will depend to a large degree on the type of trunk network employed.

4.4.5 SDS

The SDS is a service provided by specialist signal units or logistics units to physically carry messages from one location to another on the battlefield. Much of the information that needs to be transferred between headquarters would take far too long to transmit over the sparse bandwidth available in tactical communications systems. Other information is of relatively low priority and can be passed by hand, releasing communications links for more urgent traffic. Messengers can be mounted in vehicles or on motorcycles or may make use of permanently allocated aircraft on permanent allocation or other aircraft when available.

Compared to other trunk bearers, the main advantages of SDS are that it has an enormous capacity when compared to radio bearers, is relatively easy to secure as it is vulnerable only to physical threats, and is error-free (barring accidents, of course). Unfortunately, messenger services require considerable staff numbers, can take some time compared to radio transmissions, and can lack control while the messenger is in transit.

4.5 Principles of Military Communications

The doctrine of most modern armies contains a number of principles for the provision of tactical communications. These principles have been developed over many years to provide guidance for the provision of tactical communications systems to support land operations. While each principle articulates an important factor, it is rare that any one principle can be pursued vigorously without detriment to some other, and some compromise must generally be met.

Communications support the chain of command. Tactical communications systems are essential elements of a command system. The communications system must facilitate the chain of command and not constrain the ability of commanders and staffs to implement the C2 cycle. Similarly a commander's tactical plans must not be constrained by the range and capability of supporting communications systems.

Integration. A tactical communications system comprises a number of subsystems that must be integrated together efficiently. The provision of communications must therefore be controlled at the highest level of the deployed force, through the use of an integrated network management structure that supports the chain of command and can exercise tactical and technical control over the whole communications system.

Reliability. Reliability is a critical property of tactical communications systems since commanders are so heavily reliant on them and their loss is likely to cause significant tactical disadvantage. Reliability is so important on the battlefield that commanders often prefer equipment that is less capable but more reliable since it is generally much better to have a simple system that can be depended on than a sophisticated one that fails at critical points. Good planning, employment of equipment with a high mean-time-between-failure, and high standards of training reduce the risk of breakdown and increase the reliability of communications systems. System availability can also be increased by providing redundancy through the use of both standby equipments and alternative systems.

Simplicity. As tactical communications systems become more complicated and sophisticated, communications planners must resist the temptation to have equally complex plans. A simple plan is more likely to be flexible and to survive the stresses and strains placed on the network by the tactical environment. Communications equipment should also be easy to operate and simple to repair.

Capacity. A communications system must be able to cope with traffic peaks and permit all communications within the desired time frame and priorities. However, communications capacity will always be a scarce resource on the modern battlefield and measures must be devised to regulate the use of communications systems. There are three aspects of the provision of a tactical communications network with sufficient capacity: sufficient capacity of individual bearers, adequate coverage, and adequate access.

Quality. There are considerable differences in the requirements for quality (including accuracy) between the military and commercial user. At one extreme, the military user will accept intelligible voice, whereas most commercial telephone systems provide a high quality of reproduction with a very good level of speaker recognition. At the other extreme, while the requirements for errorless data transmission from terminal to terminal are the same in both environments, the commercial expectation of high-quality transmission media (error rates of certainly less than 1 in 10 [5]) may be unacceptably high in the military. This leads to difficulties when adopting commercial standards. Modern communications standards have generally been developed for transmission media such as fiber optic cables, which expect that few or no errors are introduced during transmission. Adoption of commercial data communications protocols within military networks therefore tends to require that additional error detection and correction is incorporated into the standard or especially provided by tactical communications protocols.

Flexibility. Tactical communications systems must be able to adapt quickly to any changes in the tactical environment and to continue to provide communications coverage as the deployed force maneuvers. Additionally, the same communications equipment must be able to be used for as many military tasks as possible. Equipment must therefore be flexible in the way it is deployed to meet the wide variety of tactical circumstances, as well as being able to carry different sorts of traffic such as voice, data, telegraph, and video. It is unlikely that any single communications equipment can provide all the needs of all users in all tactical situations. A communications system is flexible by providing a mix of equipment and combining the strengths of each. The main factors in the provision of flexible networks are: good planning, alternative routing, reserve equipment, personnel, and capacity on circuits, good standing operating procedures (SOP) and drills, and a high standard of training to reduce planning and deployment times.

Anticipation of requirements. The requirement for flexibility can be mitigated in some regard by anticipating the requirements of the deployed force. Commanders must therefore ensure that communications staff are kept informed throughout the C2 cycle, so that communications infrastructure can be deployed in anticipation of future plans.

Mobility. At all levels, the mobility of communications equipment must meet that of the user. For combat troops this means that radio sets must be portable or able to be fitted into fighting vehicles. For headquarters, a communications network should be able to cope with a considerable degree of movement by combat elements without needing to redeploy. The communications system must also allow commanders to command and control on the move (previously, commanders have been required to step up headquarters to achieve continuity of command). However, when required to move, the components of a communications system must have the ability to change location as rapidly as the combat elements that they serve.

Security. Due to its crucial role in support of the C2 cycle, the tactical communications network will be a prime target of adversary intelligence gathering. Protection of the information carried by the tactical communications system is therefore of prime importance. Security is a major factor in the provision of an adequate tactical communications system, and a comprehensive security architecture must be developed to provide guidance to the development of a communications architecture. There are three aspects of a secure network: physical security, personnel security, and electronic security (encryption as well as transmission and emission security).

Economy. In the commercial environment, the provision of fixed infrastructure means that there is ample bandwidth available to meet the requirements of all users. On the battlefield, communications resources will always be scarce because of the high cost of the resources themselves, as well as the personnel and materials needed to establish, operate, and maintain them. For example, while the provision of sufficient bandwidth is not generally a problem in the commercial and fixed environments, bandwidth will always be a scarce resource on the modern battlefield [16]. Communications systems must therefore be used economically and users must accept that facilities will invariably have to be shared. To maintain the principle of economy, users must ensure that: the appropriate means of communications are utilized, demands for communications are kept to a minimum, sole-user facilities are demanded only when absolutely necessary, plans are based on a realizable scale of communications, and contingency plans are available to accommodate operations if communications are disrupted for any period of time. Economy also requires individual communications equipment to be spectrally efficient to maximize the usage of limited spectrum.

Survivability.   The modern battlefield represents a harsh electromagnetic environment within which the tactical communications system must survive. A communications system is survivable if it has: sufficient capacity to handle traffic levels, an ability to manage existing capability through techniques such as dynamic bandwidth management, the necessary levels of security, low probability of intercept, resistance to jamming and interference, mobility, alternative routing, alternative means, redundancy, and sufficient reserves.

Interoperability.   The systems and networks within the tactical communications system must be interoperable with other tactical networks, strategic networks, unclassified commercial networks, as well as networks and systems of other services and allies. This interoperability is essential if information is to be able to flow seamlessly between any two points in the battlespace, and between any point in the battlespace and the strategic communications system. Additionally, new equipment will invariably need to interoperate (be backwards compatible) with current in-service equipment.

4.6   Summary

In almost all modern armies, the tactical communications system has evolved to comprise two major components: the trunk communications subsystem and the CNR subsystem. The trunk communications subsystem provides high-capacity links (terrestrial radio relay, satellite, fiber optic, or line) that interconnect headquarters at brigade level and above. The network is provided by a number of trunk nodes interconnected by trunk links to form a meshed area network. Access is normally gained through access nodes that connect to one or more trunk nodes. Voice, telegraph, data, facsimile, and video facilities are provided to staff officers and commanders. The CNR subsystem is a ruggedized, portable radio network carried as an organic communications system for combat troops (brigade level and below). Radios are invariably interconnected to form single-frequency, half-duplex, all-informed, hierarchical nets, providing commanders with effective support to command and control. In U.S. doctrine, a third subsystem, the ADDS, is provided—the utility of this third element is discussed in Chapter 5.

These subsystems have developed over time to meet the particular needs of battlefield commanders. Similarly, doctrine has evolved to articulate a number of principles of military communications. It is essential that this background is used as a starting point for an analysis of a suitable architecture for future tactical communications. While the Information Age has the promise to revolutionize warfare, it is not likely to obviate the requirement for every element of the current tactical communications system, nor invalidate all of the extant principles for military communication. Cognizant of this background, therefore, Chapter 5 conducts an appreciation of the communications support required to support tactical commanders on the digitized battlefield.

Endnotes

[1]   Further details on the history of military communications can be found in the following:

Harfield, A., *The Heliograph*, Royal Signals Museum, Blandford Camp, U.K., 1986.

Nalder, R., *The History of British Army Signals in the Second World War*, London: Royal Signals Institution, U.K., 1953.

Nalder, R., *The Royal Corps of Signals*, London: Royal Signals Institution, U.K., 1958.

Royal Signals Institution, *Through to 1970*, London: Royal Signals Institution, 1970.

Scheips, P., *Military Signal Communications, Vol. 1*, New York: Arno Press, 1980.

Scheips, P., *Military Signal Communications, Vol. 2*, New York: Arno Press, 1980.

[2]   Further details on the use of animals in military communications systems can be found in the following:

Harfield, A., *Pigeon to Packhorse*, Chippenham, U.K.: Picton Publishing Ltd, 1989.

Osman, A., and H. Osman, *Pigeons in Two World Wars*, The Racing Pigeon Publishing Co. Ltd., London, 1976.

[3]   There has been some interesting use of pigeons in more recent times—the Swiss Army only decommissioned their pigeons in 1994, and Major General John Norton, commander of the 1st U.S. Air Cavalry, initiated a small (abortive) communications experiment to revive use of carrier pigeons during the Vietnam War (see Meyer, C., *Division Level Communications*, Department of the Army, Washington, D.C., 1982, p. 45).

[4]   For more details, see Wilson, G., *The Old Telegraphs*, London: Phillimore & Co. Ltd., 1976.

[5]   The use of heliograph, signaling lamps, and signaling flags provided early commanders with useful systems for extending communications ranges. They quickly lost favor on the static battlefield of World War I, as the operator had to be exposed to obtain any reasonable ranges and they have not been used since in battlefield communications systems. It should be noted, however, that each of these forms of communications persist

in modern navies, where they are very useful for low-capacity line-of-sight communications links when the operational environment requires radio silence.

[6] FM 24-18, *Tactical Single-channel Radio Communications Techniques*, Washington, D.C.: Headquarters Department of the Army, Sept. 30, 1987.

[7] Further details on in-service CNR can be found in the following:

Jane's C4I Systems, 2000-2001, Surrey, England: Jane's Information Group, 2000.

Jane's Military Communications Systems, 2000–2001, Surrey, England: Jane's Information Group, 2000.

Witt, M., "Tactical Radios: A Never-Ending Saga," *Military Technology*, July 1999, pp. 22–32.

[8] Ryan M., *Battlefield Command Systems*, London: Brassey's, 2000, p. 139.

[9] Further information on trunk networks can be found in the following:

Blair, W., and S. Egan, "A Common Approach to Switching in Tactical Trunk Communications Systems," *Journal of Battlefield Technology*, Vol. 4, No. 2, pp. 17–21.

Hewish, M., "Tactical Area Communications Part 1: European Systems," *International Defense Review*, Vol. 5, 1990, pp. 523–526.

Hewish, M., "Tactical Area Communications Part 2: Non-European Systems," *International Defense Review*, Vol. 6, 1990, pp. 675–678.

Jane's C4I Systems, 2000–2001, Surrey, England: Jane's Information Group, 2000.

Jane's Military Communications Systems, 2000–2001, Surrey, England: Jane's Information Group, 2000.

[10] Meyer, C., *Division-Level Communications 1962–1973*, Department of the Army: Washington, D.C., 1982, p. 7.

[11] *The U.S. Army Signal Corps in Operation Desert Shield/Desert Storm*, U.S. Army Signal Center Historical Monograph Series, Office of the Command Historian, United States Army Signal Center and Fort Gordon: Fort Gordon, Georgia, 1994, p. 4.

[12] STANAG 5040, *NATO Automatic and Semi-Automatic Interfaces Between the National Switched Telecommunications Systems of the Combat Zone and Between these Systems and the NATO Integrated Communications System (NICS)–Period From 1979 to the 1990s*, NATO, Brussels, Oct. 23, 1985.

[13] EUROCOM D/1 Standard, *Tactical Communications Systems—Basic Parameters*, NATO, Brussels, 1986.

[14] STANAG 4208, *The NATO Multi-Channel Tactical Digital Gateway—Signalling Standards*, NATO, Brussels, Nov. 15, 1993.

[15] STANAG 5046, *The NATO Military Communications Directory System*, NATO, Brussels, Aug. 8, 1995.

[16] As described in Chapter 5, this bandwidth paucity on the battlefield is mainly caused by the limitations of the physics associated with the methods of propagation available to tactical communications systems.

5

A Communications Architecture for the Digitized Battlefield

5.1 Introduction

Current tactical communications systems have evolved to meet users' needs as the conduct of warfare has changed, particularly over the last several hundred years. However, if land warfare is to be revolutionized by Information-Age technologies, the tactical communications architecture must become an integral part of a force's ability to prosecute war. While the success of a commander has always been contingent upon the provision of reliable communications and information systems, the success of a modern commander relies heavily on the domination of the electromagnetic spectrum. This critical interdependency between communications and command and control requires a reconsideration of an appropriate architecture for tactical communications systems.

This chapter develops an architectural framework [1] to define the tactical communications system. It begins by outlining key design drivers that shape the architecture of a tactical communications system. Options for a mobile tactical communications system are then examined and a suitable framework is developed within which architectural issues may be subsequently considered.

5.2 Design Drivers

The following issues represent key design drivers for the development of architecture for the tactical communications system.

5.2.1 Principles of Military Communications

As described in Chapter 4, the doctrine of most modern armies is based on a number of principles for the provision of tactical communications. Briefly, the major principles are: *communications support the chain of command, integration, reliability, simplicity, capacity, quality, flexibility, anticipation of requirements, mobility, security, economy, survivability,* and *interoperability.* These principles have been developed over many years to provide guidance for the provision of tactical communications systems to support land operations. While the modern digitized battlefield is different in many respects to those of the past, the principles of communications are verities of warfare that must also be considered as design drivers for a tactical communications architecture.

5.2.2 Size of the Supported Force

At the strategic and operational levels, the provision of high-capacity communications is relatively straightforward and can quickly be implemented using commercial technologies and networks. Headquarters at this level are relatively static and can be well served by high-capacity systems such as satellite and PSTN. Tactical elements are not so well served by commercial technologies, however, and military tactical communications systems suffer from the last-mile difficulties that plague commercial networks.

In the analysis that follows, therefore, attention is focused at the level of the division-sized group, that is, on division and below. While the architecture of the tactical communications system is developed with cognizance of larger formations, the most difficult problem is the provision of sufficient communications capacity to tactical combat forces at division and below.

5.2.3 Communications Support for the Spectrum of Operations

The communications architecture must be able to support a force in a range of operational deployments. The spectrum of operations varies from conventional, high-density motorized or mechanized operations to low-density operations, or in peacekeeping or peace-enforcing operations. The tactical communications system must therefore provide similar interfaces, regardless of the type of operation. This concept produces two main design drivers:

1. User equipment must be similar in all types of operations. The user should not be expected to have to adjust to a different communications interface in different operational circumstances. This means that command posts, vehicles, and weapons platforms must be configured with a flexible internal communications network that can interface to the range of available communications systems.

2. Specialist communications equipment must be similar in all types of operations. All specialist signal units should therefore be similarly equipped. While the vehicles may differ between units that support motorized, mechanized, and armored forces, the basic equipment configurations should remain similar, which will greatly ease procurement, logistic support, and training.

5.2.4 Command and Control on the Move

On the future battlefield it is essential that commanders can command and control on the move. A commander should be able to control force elements, regardless of location. This includes relatively simple activities such as maintaining continuity of command and control while deploying from the barracks, as well as the much more difficult task of staying in contact while moving around the battlespace in a variety of tactical environments. The requirement for command and control on the move provides one of the most significant design drivers because it impacts on the power available for transmitters and the antenna size that can be supported. These issues are discussed in more detail in Section 5.3.

5.2.5 Communications Support Situational Awareness at All Levels

Communications must support *situational awareness* to commanders at all levels (in real time or near real time) to provide accurate knowledge of adversary, friendly, neutral, and noncombatant entities. Situational awareness is the knowledge of the operational environment required to gain the level of understanding necessary to achieve decisional superiority over an adversary. Most modern commanders would expect near-real-time situational awareness at their headquarters, which provides a significant design driver for the tactical communications system because it implies a battlefield-wide network that can support the transfer of information from any point to any other point.

Traditional hierarchical communications architectures are not well suited to the provision of a modern battlefield network, particularly for the transfer of data. To illustrate this point, consider a notional mechanized brigade, the basis of which is a mechanized platoon that has four vehicles. Real-time situational awareness would require that the location of every vehicle is known at any instant. This is an unrealistic expectation. Near-real-time situational awareness requires that the location of each vehicle be reported to an accuracy within at least 10m (for battlefield identification and to prevent significant errors on sensor-to-shooter target sharing). A location report is likely to require at least 100 bytes (800 bits) of data, which, with the added overhead of EDC, would be at least 200 bytes.

Assuming that the platoon is advancing at its maximum cross-country speed of 40 km/hr [11 meters per second (m/s)], each vehicle must transmit a location report at least every second (at least) so that its location is known within 10m. The platoon net would have therefore have to have sufficient capacity to cope with $4 \times 200 \times 8 = 6.4$ Kbps, the company net would require capacity for 25.6 Kbps, the battalion net would require 102.4 Kbps, the brigade net would require approximately 500 Kbps, and the divisional net would require at least 2 Mbps.

However, the current tactical communications systems are poorly placed to provide the capacity required. Figure 5.1 illustrates the problem. At the platoon level, vehicles will generally be in sight of each other and can use UHF communications, which can provide ample capacity for the needs at this level. However, between platoon and company, VHF radio is required; between company and battalion, VHF radio is required, with HF required for longer ranges; and between battalion and brigade, VHF radio normally has insufficient range and HF radio is used. Currently, the throughput of most analog CNR is restricted to approximately 8 Kbps at VHF and 300 bps at HF.

Thus, the data rate required increases as the situational awareness data is passed back to brigade and divisional headquarters. Unfortunately, in most dispersed deployments, the available capacity is significantly smaller between higher headquarters than between lower command elements.

To avoid the aggregation of data as it is passed up the chain of command, CNR nets must change in architecture from a hierarchical to a networked structure. Even so, the constraints of physics mean that there will not be sufficient capacity available for more than low-capacity data rates (i.e., low by fixed network standards). While future technological advances may allow

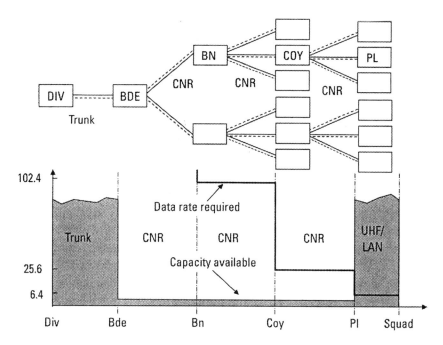

Figure 5.1 Data rate required versus capacity available.

some increase in data rates across CNR, it is unlikely that such nets will ever carry the volume of traffic required to support situational awareness. This is particularly true when it is recognized that these CNR nets will also be required to carry voice data, significantly reducing the bandwidth available for data. It should also be noted that information will need to be passed from higher to lower levels, making the situation even worse.

Clearly, traditional hierarchical architectures are unsuited to the provision of a battlefield network across which users can share large amounts of data. Additionally, if the data rates necessary for near-real-time situational awareness are to be provided, the CNR and trunk subsystems are unable to cope and a higher capacity network that is dedicated to the transfer of situational awareness data must be provided.

5.2.6 Seamless Connectivity

If concepts such as network-centric warfare are to be implemented on the future battlefield, the communications architecture must support seamless connectivity between any two points in the battlespace, and between any point in the battlespace and any point in the strategic communications system. While it will not always be desirable to support such extremes of connectivity, the possibility of such connections must be supported in the architecture if the full power of network-centric warfare is to be realized.

5.2.7 Organic, Minimum-Essential Communications

The deployed force must have an organic, field-deployable tactical communications system that meets minimum essential requirements for communications to support command and control. This minimum-essential system must provide guaranteed, robust, flexible communications that support the force whether deployed on foot, motorized, mechanized, or armored. An organic tactical communications system is part of the force's combat power. Since the commander owns the minimum-capacity network, the essential communications requirements of the force can be guaranteed in any deployment. This also requires that the network is modular so that units and subunits are self-sufficient in communications functionality when deployed separately from the main force.

5.2.8 Expandable Communications

Since its organic communications will invariably be limited, the deployed force must be able to make use of other battlefield and strategic communications systems when available. While essential requirements are provided by organic communications, additional capacity, redundancy, and reliability can be provided by using overlaid communications systems such as the commercial PSTN, satellite communications, and theater broadcast. These systems must be seamlessly integrated with the tactical communications system.

5.2.9 Scalable Communications

Within all available assets the brigade must have the ability to provide scalable communications. That is, a small advance party must be able to deploy taking with it sufficient communications for its task. As the force builds, the communications system must be able to grow to accommodate the size of the force.

The tactical communications system must be able to support the force when it is deployed in any one of its roles. In the extreme, the communications system must be able to provide communications in conventional high-density deployments where brigades might be deployed with a 25-km frontage, as well as support widely dispersed deployments where a single brigade may be spread over an area with a 500-km radius.

The second aspect of range is the ability to reach back from the tactical communications system to the strategic communications system. The tactical communications system must therefore be able to support high-capacity communications from the area of operations back into the strategic environment, which may require communications around the world.

5.2.10 QoS

The requirement to support both voice and data communications on the battlefield is often characterized as a voice-versus-data debate, which tends to belie the fact that both have relative advantages. Voice has long been the preferred means of communications on the battlefield because:

- Voice conveys the imperative of the situation.
- Voice carries the personality of the speaker.
- Voice is user-friendly in that it is a familiar interface to the user, allows for conversational interchange, only requires the use of one hand, and can be conducted while the user is running, bouncing around in a vehicle, or stationary.
- Voice does not require off-line preparation.
- Voice is interactive and immediate.

However, data messaging has a number of considerable advantages:

- Terminals do not need to be manned permanently.
- Messages can be acknowledged automatically, preformatted, and prepared off-line.
- EDC techniques ensure that messages have more chance of being received correctly under poor telecommunications conditions.
- Encoding in digital form allows the efficient transfer of information through networking, particularly the integration of networks to allow the seamless transfer of information between any two points in the battlespace.
- The bandwidth required for a message is much smaller when passed as a data text message compared to a digitized voice message. The reduced transmission time also assists in reducing the probability of detection.

While warfare ever remains a human endeavor, commanders will want to communicate using voice. However, concepts such as network-centric warfare cannot be force multipliers unless the communications architecture can support the rapid transfer of data across the battlefield. Since the tactical communications system must serve both types of users, both voice and data communications must be supported.

In addition to data messaging, battlefield entities need to transfer other forms of data, including video and database transfers. With the proliferation of battlefield information systems, these types of data have the potential to dominate over the more traditional data messaging. In fact, since most modern battlefield communications systems are increasingly digital, all forms of traffic are invariably digitized before transmission. Since there are many different types of data on the modern battlefield, it does not necessarily make sense to distinguish them by their source. Rather, it is more useful to characterize them by the requirements that each type of data has for services across the network. In that regard, there are two main types of traffic: those that require real-time services (predominantly video and voice) and those that require non-real-time services (computer-to-computer transfer).

5.2.11 Low Probability of Detection

Low probability of detection (LPD) is critical to tactical communications systems. Survival on the modern battlefield requires the protection of communications systems as the first step in protecting the command systems that they support. LPD techniques include short-duration transmission, spread spectrum (direct sequence spread spectrum as well as frequency hopping), directional antennas, low power settings, terrain screening, and the use of airborne relays so that ground terminals can direct their power upward, away from a land-based adversary.

5.2.12 Jamming Resistance

Jamming resistance is also critical to the protection of communications links. Operation in a harsh electromagnetic environment requires the ability to implement measures to provide resistance to jamming. Techniques listed in Section 5.2.11 for LPD are also relevant to increasing jamming resistance. Other techniques that may conflict with LPD requirements include increased power, strong error coding, jamming-resistant modulation, and adaptive antennas with steerable nulls.

5.2.13 Precedence and Preemption

The requirements for precedence and preemption are unique to military communications systems. Precedence implies some priority to a conversation or message. Originally a service provided to commanders in the circuit-switched networks, the concept extends to packet-switched networks where some data has a higher priority because of its information content rather than its QoS requirements. To ensure that higher precedence communication can occur, lower priority traffic can be preempted. That is, lower priority users may have communications terminated or delayed to make way for higher precedence users.

These requirements do not exist in commercial networks, which are more egalitarian (although parental precedence over teenagers and modems may be desirable in many homes) in both voice and data networks. Precedence and preemption are essential for tactical communications systems, however, if sparse bandwidth is to be used effectively. This places significant constraints on the military use of commercial systems where military users must compete on an equal footing with civilian users (and possibly even an adversary) for access to channels.

5.2.14 Electromagnetic Compatibility

Mobile users need to be connected to the network by a radio link. A large number of mobile users therefore requires a large number of radio channels, which leads to difficulties in frequency allocation and management, as well as cositing difficulties due to the installation of equipment in close vicinity within a vehicle. Electromagnetic compatibility (EMC) is therefore an important consideration in equipment specification, design, and operation.

Military forces are not alone in this regard and EMC is an important consideration in commercial networks. However, the tactical communications system provides perhaps the harshest environment due to the large number of equipment and users in a small area. For example, a single U.S. heavy division may have up to 10,700 RF emitters in an area of approximately 45 km by 70 km [2].

5.2.15 Supported Systems

The tactical communications system architecture must support a wide variety of battlefield, joint, and combined systems, including command elements, maneuver elements, logistic elements, sensors, weapons platforms, information systems, information services, and network management. Each of these elements is discussed in more detail in Section 5.4.

5.2.16 Operation in All Geographic and Climatic Conditions

Tactical communications equipment is required to operate in a range of conditions beyond those normally expected for commercial equipment, including immersion in water, a temperature range of −20°C to +70°C, and to withstand extremes of vibration, shock, pressure, and humidity. These requirements relate to individual equipment but can be reduced, however, if equipment can be protected in other ways. For example, vehicle-mounted equipment may be able to be protected at the vehicle-system level; commonly, the operating temperature range may be reduced if air conditioning is provided within the vehicle.

5.2.17 Range, Capacity, and Mobility Trade-off

The tactical communications system must be able to support the force when it is deployed in any one of its roles. In the extreme, the communications system must be able to provide communications in conventional high-density deployments where brigades might be deployed with a 25-km frontage, as well as support widely dispersed deployments where a single brigade may be spread over an area with a 500-km radius.

The second aspect of range is the ability to reach back from the tactical communications system to the strategic communications system. The tactical communications system must therefore be able to support high-capacity communications from the area of operations back into the strategic environment, which may require communications around the world.

Communications systems provide ranges that vary from several meters to thousands of kilometers, with capacities from hundreds of bits per second to more than 1 Gbps, and with degrees of mobility varying from hand-portable systems that are operable on the move to large fixed installations that must be shut down before moving. An ideal communications system would provide long range, high capacity, and high mobility. However, as discussed in Chapter 4, the delivery of all three of these characteristics is problematic, and a communications system with ground-based terminals may be able to exhibit two of them, but not all three at once.

High-capacity communications.   The capacity of a communications channel depends on the bandwidth of the channel and the signal-to-noise ratio at the receiver. Channel capacity can be increased, therefore, either by increasing the channel bandwidth or increasing the transmit power. High capacity can be achieved by the use of a cable, optical fiber, light through air (usually a highly directional beam generated by a laser), or wireless RF connection operating in the VHF or higher frequency band. RF propagation in the VHF and higher bands is by direct wave (implying line-of-sight). Over-the-horizon ranges can be achieved by scattered-wave communications. High-capacity channels are not usually available in the HF and lower frequency bands due to the limited bandwidth available, although HF has also been used to provide overflow capacity when VHF nets became overcrowded [3].

Long-range communications.   Communications range is a function of the frequency of operation and the heights of the transmit and receive antennas. Long range can be achieved by using a frequency in the HF or lower band, utilizing surface-wave or sky-wave propagation, by using either high transmit power and an elevated antenna or one or more (probably elevated) relays with direct-wave propagation at VHF or higher frequencies, by using scattered-wave communications, or by using a cable or optical fiber, possibly involving one or more relays.

High-mobility communications.   Mobile forces require communications systems that are based on RF wireless communications, that is, high mobility cannot be achieved using either cable or optical fiber. High-mobility wireless communications requires the use of small antennas. It is not possible to use large antennas because they require support from a mast (as in the case of efficient sky-wave antennas) or elevation on a mast (for long-range line-of-sight communications or scattered-wave communications). Using currently available technology, high mobility is normally only possible with omnidirectional antennas, although further development of antenna arrays may make ground-based, directional antennas feasible with a limited capability for communications on the move. It is unlikely in the short term, however, that technology will be available to enable directional antennas to be used from a ground vehicle while it is moving, particularly cross-country. Man-portable communications systems also impose restrictions on transmit power in order to extend battery life.

High-capacity/long-range communications.   High capacity and long range can be achieved by the use of cable or optical fiber, scattered-wave wireless communications, or line-of-sight wireless communications with directional antennas or high transmit powers. All of these solutions limit mobility.

High-capacity/high-mobility communications.   High capacity and high mobility can be achieved by line-of-sight, wireless communications using low, omnidirectional antennas. Range is limited by the line-of-sight available from low antennas. For man-portable communications, range is also limited by the use of low transmit powers.

Long-range/high mobility communications.   Long range and high mobility are achieved by the use of surface-wave wireless communications or by the use of an elevated relay with line-of-sight wireless communications. The capacity of surface-wave wireless communications is limited by the small bandwidth available. The capacity of long-range, highly mobile line-of-sight communications systems is limited by the use of omnidirectional antennas and the limitations on transmit power imposed by the requirement for mobility. This trade-off between range, capacity, and mobility is illustrated in Figure 5.2. Also shown is the trade-off choice commonly made for a number of communications systems, which are described in further detail next.

Optical fiber.   Optical fiber can provide capacities in excess of 100 Gbps over ranges of many kilometers. It is time-consuming, however, to lay and recover, allowing network terminals essentially no mobility during communications. In tactical communications systems the utility of optical fiber is therefore limited to providing high-capacity links within vehicles and command posts. In some cases, for example, when at the halt, vehicles can be interconnected with fiber to create a high-capacity LAN. Optical fiber would usually be avoided on longer links due to the long setup and teardown times required.

WLAN.   WLAN technologies (such as IEEE 802.11 [4]), including personal network technologies (such as Bluetooth [5]), provide full operation on the move and very high capacities (up to 1 Mbps for Bluetooth and 54 Mbps for IEEE 802.11). These high capacities and degrees of mobility are provided at the expense of very short ranges (on the order of tens of meters). Again, the uses of wireless LANs in tactical communications systems are limited to the provision of high-capacity LAN. Additionally, the technologies themselves are not scalable and are therefore not suitable for creating WANs.

Trunk communications systems.   For terrestrial radio-relay links, long range (in excess of 10 km where line-of-sight exists) and high capacity (up to approximately 2 Mbps) are provided by the use of elevated directional antennas, usually mounted on masts at least 10-m tall, and by the use of transmit powers up to approximately 100W.

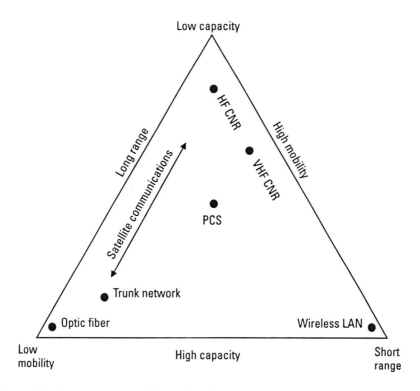

Figure 5.2 Range-capacity-mobility trade-off.

CNR.   The main role of CNR is to provide communications for the command and control of combat troops, which requires high mobility, with full operation on the move. VHF CNR, therefore, sacrifices capacity compared to terrestrial radio relay to provide moderate range with high mobility. HF CNR can provide longer range while still maintaining mobility, through the use of surface-wave propagation. The capacity of current analog VHF CNR (usually 16 Kbps per channel) is higher than HF CNR (usually less than 3 Kbps per channel) because of the larger channel bandwidth used for VHF CNR (25 kHz for VHF versus 3 kHz for HF).

Cellular telephones.   Cellular telephone systems provide moderate capacity and a range (between mobile terminal and base station) of up to several tens of kilometers. The mobility of the mobile terminal is very high, while the base station is invariably fixed with antennas mounted on hills, tall masts, or buildings. Directional antennas are usually employed at the base station to reduce intercell interference. The effective channel bandwidth for most second-generation cellular systems (such as GSM) is 25 to 30 kHz for each uplink and downlink. Data capacities are approximately 10 Kbps per channel.

Satellite communications systems.   Satellite communications systems are designed to provide long-range communications. Systems vary, however, in their choice of the trade-off between capacity and mobility. Small, handheld terminals with omnidirectional antennas (such as an Iridium handset) provide a capacity of approximately 2.4 Kbps, a military tactical satellite system with a 2.4-m parabolic antenna mounted on the back of a truck may provide up to approximately 2 Mbps, while a large, fixed ground station whose highly directional, parabolic antenna may have a diameter in excess of 60m may be capable of more than 10 Mbps.

5.3   Options for a Mobile Tactical Communications System

As discussed in Chapter 4 and illustrated in Figure 5.3, the current tactical communications system has evolved to comprise two major components. The trunk communications system provides high capacity links (using radio relay, satellite, optical fiber, or line bearers) that interconnect headquarters at brigade level and above. The CNR subsystem is a ruggedized, portable radio network carried as an organic communications system for combat troops (at brigade level and below).

There are a number of major problems with the current tactical communications system if it is to support command and control in future land warfare:

- Due to its hierarchical, analog voice net structure, the CNR subsystem is poorly placed to provide a network to transfer data between combat units.

- The networks are not seamlessly integrated to allow the transfer of information between any two points in the battlespace.

- Information cannot be passed between CNR nets, or between the CNR and trunk subsystems (except for the rudimentary interface of CNRI).

- The tactical communications system is not seamlessly integrated with the strategic communications system to allow the transfer of

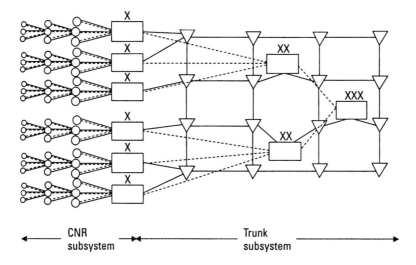

Figure 5.3 The current tactical communications system.

information between any point in the battlespace and any point in the strategic communications system.

- There is not sufficient capacity below brigade to cope with future levels of data traffic required to provide commanders with near-real-time situational awareness.

- Neither the current CNR subsystem nor the trunk communications subsystem provides enough range to allow the dispersal of divisional elements when required to meet the tactical situation.

Philosophically, support for command and control in future land warfare requires the tactical communications system to be a single logical network (as shown in Figure 5.4) to provide connectivity between any two points on the battlefield. The tactical communications system is an organic asset that provides the minimum essential voice and data communications requirements to support situational awareness within the brigade and to allow for the transfer of command and control information. An interface is required with the strategic communications system to provide seamless connectivity between any two points in the battlespace and between any point in the battlespace and any point in the strategic communications system.

The development of a suitable architecture for the tactical communications system can draw on the considerable body of knowledge available in

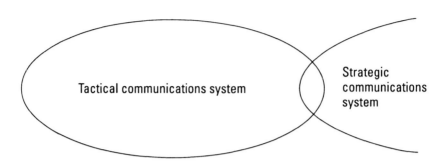

Figure 5.4 Tactical communications system and strategic communications system.

existing commercial and military networks that provide mobile communications. However, in some respects, *mobile communications* is a misnomer when used in the commercial environment because only the user is mobile in such systems; the vast majority of the communications system (the communications network) is very much fixed, with mobile access to this fixed infrastructure being provided by a wireless connection. In the military environment, the provision of a mobile communications system normally implies that both the user and the network infrastructure are mobile. Mobility therefore means markedly different things in the commercial and military environments. Consequently, while many commercial communications technologies are useful in the military environment, the mobility of the network infrastructure for military communications systems tends to require unique solutions.

Even within military mobile networks, a distinction must be made between networks provided with mobile infrastructure and transportable infrastructure. It is therefore useful to consider potential mobile network architectures under the following three categories, not necessarily in terms of the mobility of the network, but rather in terms of the mobility of the network infrastructure: *fixed network infrastructure, semimobile network infrastructure,* and *mobile network infrastructure* [6]. Sections 5.3.1 to 5.3.3 briefly examine the commercial and military mobile communications systems available within each of these categories.

5.3.1 Fixed Network Infrastructure

Commercial mobile communications systems are based on a fixed network infrastructure with mobility provided by a wireless interface between the user and the network. Major systems include terrestrial mobile telephony and wireless networking.

5.3.1.1   Terrestrial Mobile Telephony

Mobile telephone systems provide wireless access to users who can have varying degrees of mobility. There are two main types of mobile phone systems: cellular and cordless phones.

Cellular phones.   In cellular phone systems, mobile telephone users are connected by wireless telephone handsets to one of a number of base stations, each of which serves a number of mobile users within an area called a cell. Bases stations are then interconnected using high-speed trunk lines provided by fixed infrastructure, normally that of the PSTN. A group of base stations is controlled by a mobile switching center, which supports the handover of a mobile user as that user moves from one cell (base station) to another. Users are therefore able to make calls as they roam throughout the network. Cellular telephone networks are designed predominantly for voice, but can also pass data at rates of up to 9.6 Kbps (although the introduction of third-generation systems promises to increase data rates considerably).

Cordless phones.   A number of applications do not require the complexity of cellular telephone systems, particularly the arduous (and expensive) requirement to hand over users. These simpler mobile telephone techniques (e.g., CT2 and DECT) allow a mobile user to be supported by a single base station for the period of a call. Handover is generally not supported, although as an exception to this, DECT does support a limited form of handover. Range is therefore limited to line-of-sight between the user and the base station. Cordless phone systems are generally designed for voice calls only.

5.3.1.2   Wireless Networking

WLANs provide some flexibility for computer users who want to be able to detach and attach quickly to a network, or for networks—such as those provided in classrooms—where users move in and out of the network on a regular basis. WLANs can be based on IR or RF bearers. IR WLANs have very limited range, since they are constrained by line-of-sight and high atmospheric attenuation. RF WLANs have longer ranges but are still too short for wide-area coverage. While wireless networking has great utility within command posts and other locations where computers are deployed on the battlefield, ranges are too limited to participate in the broad-area coverage required of the tactical communications system.

5.3.1.3 Military Utility

Although they may have the potential to provide the highest capacity and QoS to deployed forces, fixed-infrastructure (commercial) mobile communications systems have limited utility as the basis for the tactical communications system:

- *Provision of infrastructure.* In mobile communications systems, only the user is mobile; the considerable network infrastructure is fixed. Terrestrial mobile communications systems require the provision of a significant amount of infrastructure in the form of base stations every 20 or 30 km. In addition, base stations are connected by the fixed infrastructure of the public telephone network. Since the provision of this infrastructure cannot be guaranteed within any potential area of operations, the necessary equipment and personnel would need to be provided by the tactical communications system. Significant infrastructure would be required to support the force over a widely dispersed area.

- *Range.* Calls in terrestrial mobile communications systems cannot be placed outside the range of the network—not only are communications ranges to base stations limited by terrain but in systems such as Global System for Mobile Communication (GSM), range is limited to 35 km due to the power balancing required at the base station.

- *Flexibility.* Since they are connection-oriented, mobile cellular communications systems lack flexibility. A user can only make a call to another user, while some systems may support, or may be able to be adapted to support, broadcast or conference calls. All-informed communications are therefore not easily obtained. In addition, while the service provider can provide connection to a data network, it is very difficult to provide the type of network required to support the seamless transfer of data between any two points in the battlespace, since any such network would be essentially circuit-switched. This is adequate if the mobile user is the initiator and the connection is simply to a server—problems ensue if contact with the mobile user is sought from a server on the fixed network (or another mobile user through the fixed network).

- *Capacity.* Cellular telephony systems have limited capacity—currently about 9.6 Kbps. While there is promise that these rates will increase somewhat in the near-to-medium term [7], they will still only serve single military users because only a single connection is provided. It should also be noted that higher-capacity cellular systems are possible with a limitation on the number of simultaneous active connections.

- *Cost-effectiveness.* Significant quantities of infrastructure are required to provide base stations that cover a population of users. First, in regions that have high populations, a large number of base stations are required to accommodate possible calls. Then, in regions of low population, a significant number of base stations are still required as communication is terrain-limited to tens of kilometers. The provision of this infrastructure on the battlefield is not cost-effective to provide complete coverage.

- *Security.* Both terrestrial mobile telephony and wireless networking are very limited in their ability to provide security services to applications. There are few mechanisms for voice security, and those digital networks that do provide encryption use short keys that can be easily broken.

- *Precedence and preemption.* Commercial systems do not recognize that there are users with different precedence and military users will have to compete for channels with commercial users and perhaps even adversaries.

- *Mobility.* Commercial PCS systems support mobile users with small handsets. High-capacity data systems such as Teledesic require larger terminals that are semimobile.

- *Resistance to jamming.* PCS systems do not support any techniques that would provide resistance to jamming. Individual cells are therefore very vulnerable, as the base station can be attacked easily by physical and electronic means.

These limitations mean that fixed-infrastructure systems have little utility for inclusion in the tactical communications systems because they cannot provide the minimum, organic, flexible communications required to support voice requirements. Additionally, such systems do not easily support the provision of a data network. However, mobile telephone systems provide a very useful adjunct to the tactical communications systems and must be able to be used when available.

5.3.2   Semimobile Network Infrastructure

In these networks, mobile users connect to semimobile infrastructure, which can deploy and redeploy to meet changes in the operational requirement. Some degree of infrastructure mobility is essential to support military operations.

5.3.2.1   Trunk Communications Networks

Most modern armies currently employ trunk communications systems that provide semimobile infrastructure to support communications between headquarters. The mobility of these trunk networks is constrained by their size and deployment times. However, they are sufficiently mobile to allow combat forces, and particularly command elements, to be fully mobile within the area of operations. In particular, trunk communications networks provide high-capacity point-to-point links between transportable switching nodes. The network deploys to cover an area of operations and command elements connect into a convenient node.

5.3.2.2   Cellular Communications Systems

As outlined in Section 5.3.1, mobile communications systems tend to rely on too much fixed infrastructure to be of any great use in a tactical communications system. However, with reduced functionality, a mobile base station could be deployed to support isolated communities of users, such as in command posts, logistics installations, and airfields. If the tactical situation allows, mobile base stations could be interconnected by high capacity radio links. This becomes particularly feasible when base stations are collocated with (or integrated into) trunk network nodes. In fact, SCRA provides such a service and has been in service with trunk networks since the mid-1980s, quite some time before cellular communications systems became as readily available as they are today.

5.3.2.3   Military Utility

While modern trunk communications systems may one day become fully mobile, they are currently constrained in their mobility by their size, power requirements, and the need to orientate the high-gain antennas required to support high data rates over useful distances. Some research is currently being conducted in this area, but planned high-capacity radios (45–155 Mbps) have very limited ranges (tens of kilometers only). Current trunk communications systems architectures are therefore likely to remain for the next 10 to 15 years.

Additional systems, such as semimobile cellular communications systems, will continue to find applications as adjuncts to trunk communications systems. However, they are unlikely to develop the functionality to replace them.

Therefore, the high-capacity backbone of the tactical communications system must still be provided by a system similar in design to current modern trunk communications systems. A number of changes are required, however. In particular, trunk communications must be extended to below brigade headquarters and the trunk network must be seamlessly integrated with other battlefield networks.

5.3.3 Mobile Network Infrastructure

Mobile networks provide the greatest flexibility and mobility to support military operations—all users are mobile, as is the network itself. These types of networks, apart from satellite communications systems, tend to be unique to the military environment.

5.3.3.1 CNR

CNR is the traditional means of providing communications on the move. Small, robust radios are combined into a flexible tactical system providing single-frequency, half-duplex, all-informed communications in support of the command and control of combat troops. Radios are mounted in vehicles or are carried in soldiers' packs. Radios provide both user terminals and network nodes. Mostly voice communications have been provided, although data communications are increasingly available in modern radios. The major disadvantage of CNR is that it is generally terrain-limited, rather than power-limited, which causes additional difficulties in establishing and maintaining communications.

5.3.3.2 Packet Radio

Packet radio systems were developed as an extension of CNR systems. Their main design driver was to be able to handle the requirement to send data over tactical mobile links. Radios are digital and exchange information by breaking messages up into packets and then routing them around the network. Packets are stored at each radio and then forwarded when the next link is available. Packets may take a number of hops to reach their destination. While packet radio systems can have fixed infrastructure, they are most useful for military use when mobile. Packet radio networks provide one of the few architectures available for providing a data network on the battlefield that could support mobile users with mobile network infrastructure. Like CNR, however, they are terrain-limited, which causes difficulty in maintaining network connectivity between mobile forces that naturally keep low in the terrain.

5.3.3.3   Satellite Communications

Mobile communications can also be provided by satellite communications systems. While the network infrastructure is not mobile in the sense that we have considered other infrastructure, it is generally ubiquitous over all possible areas of operations. Since the infrastructure itself does not place any constraints on the mobility of the users, satellite communications have been included in this mobile-infrastructure category.

GEO communications systems.   GEO satellites orbit the Earth at approximately 36,000 km above the Earth's surface. Because of this long range, mobile users need to have reasonably large antennas, and the phones are considerably larger than cellular phones (normally small briefcase-sized terminals). Users connect to a terrestrial gateway that is connected to the PSTN. If one mobile satellite user wants to talk to another, connection is made through the terrestrial gateway, requiring two uplinks and two downlinks. However, the user is provided much greater mobility than in terrestrial cellular networks, since the coverage of the system is far greater. It should be noted that the use of UHF satellite-based CNR systems may simplify communications by using bent-pipe architectures.

LEO communications systems.   LEO communications satellites have a much lower orbit than GEO—approximately 800 km above the Earth's surface. LEO mobile communications systems therefore require much smaller terminals (slightly larger than a modern cellular phone) than GEO systems. In systems such as Iridium, users communicate directly to each other without using terrestrial infrastructure (after gaining approval from a terrestrial gateway), supported by intersatellite links. Other systems such as Globalstar provide communications between terminals by switching through base stations on the ground. If a sufficient number of satellites has been provided, LEO-based mobile communications system potentially covers the entire surface of the Earth, providing users with complete mobility and the ability to make a call anywhere at any time.

In the near future, satellite-based mobile communications systems are likely to be integrated seamlessly with terrestrial cellular systems through the work being conducted in the development of third-generation mobile communications systems.

5.3.3.4 Fully Meshed and Repeater-Based Networks

Fully meshed architectures such as Joint Tactical Information Distribution System (JTIDS) (see Section 8.4.4.1) have been developed for air-to-air and air-to-ground communication providing up to 30 nets, each of which is shared on a TDMA basis. Communications are broadcast to the net providing considerable survivability since there are no critical nodes. Communications are line-of-sight, although JTIDS has a relay capability to support communications beyond line-of-sight. However, setting up a relay requires manual configuration by an operator and the use of relays also significantly degrades overall system performance.

Fully meshed networks are inherently inefficient, however, because each transmitter can make range (timing) and Doppler corrections for only one receiver in the network (or rather, it can only make the same corrections for all receivers). The network is inefficient, therefore, because sufficient guard time has to be allowed for stations at the full extremity of the network (300–500 nm for JTIDS). Guard times can be much smaller in repeater networks because transmitters can adjust their transmit times to correct for range differences to the repeater, and guard times need only be as long as required to accommodate the uncertainty in propagation time to the repeater (plus timing errors).

Additionally, meshed networks cannot reuse frequencies because all nodes are considered to belong to the one community. If more than one base station is used in a repeater-based network, frequencies can be reused, allowing for significant increases in capacity over fully meshed and single-repeater networks. Some terrestrial networks such as Enhanced Position Locating and Reporting System (EPLRS) provide a compromise solution with limited meshed networks controlled by a base station. EPLRS stations are also capable of automatic repeating between stations.

5.3.3.5 Airborne Repeater

An airborne repeater would be particularly useful because it would be able to accommodate a user community across a much larger area than a terrestrial repeater. With accurate position and time information available at each node through GPS, it should be possible to virtually eliminate guard-time overhead. Using multiple-access protocols to an airborne repeater would allow high (UHF or X-band) frequencies to be used providing up to 500 MHz of spectrum. The net effect of these factors might be a hundred-fold increase in system capacity relative to a system such as JTIDS. With multiple relays and frequency or code reuse, even larger capacities might be realized [6].

Additional advantages would accrue if the airborne platform was to be used for other communications functions, such as an airborne cellular base station, HF, VHF, or UHF rebroadcast, UHF theater broadcast, or even as a surrogate satellite. In other words, the platform could be used with existing terrestrial terminals to extend their limited ranges. An airborne repeater therefore has the advantages of providing range extension to terrain-limited nets as well as opening up new opportunities in multiple-access communications that can provide coverage of the entire area of operations. These issues are discussed in more detail in Chapter 9.

5.3.3.6   Military Utility

While all of the previously mentioned communications systems have the potential to provide tactical mobile communications architecture, the following apply.

Architecture.   Satellite-based PCS communications systems are point-to-point and are therefore not well suited to all-informed voice nets or data networks providing seamless connectivity. They are, however, ideally suited to employment as an overlaid communications system.

Range.   CNR and packet radio solutions are difficult to provide and maintain because the radios are terrain-limited in range, rather than power-limited. This is a function of low antennas mounted on low-profile vehicles or on soldiers' backs. Range limitations are even worse for fully meshed systems such as JTIDS, where it is unlikely that one terminal can see any more than a few others at any one time when deployed in ground-based units. Due to their significantly greater heights, satellite and airborne systems are able to cover the area of operations and be visible to all ground terminals. UHF CNR can have very long-range extensions offered by satellite-based systems.

Capacity.   CNR and packet radio solutions have capacities limited by the transmission techniques and modulation schemes as well as the possible multiple access techniques. GEO systems are limited in capacity by the size of antenna that can be mounted on the mobile platform—significant rates can only be sustained by semimobile terminals. Mobile terminals can be used with LEO systems, but data rates are constrained to 9.6 Kbps in the near term. UHF satellite communications also have limited capacity, particularly for a reasonably large number of terminals.

Cost-effectiveness.   CNR and packet radio systems provide cost-effective solutions to the voice communications requirements of combat troops since they provide relatively cheap, flexible communications to support the full range of deployments. Data rates are generally very low, however, and only able to support limited transfer of situational awareness data. Unfortunately, the cost of satellite PCS systems tends to preclude their widespread use, although they are very effective forms of communications for limited use by reconnaissance teams and advance parties. Airborne systems provide high-capacity, long-range, flexible communications at some significant cost (although aerostat solutions are an order of magnitude less than uninhabited aerial vehicles (UAVs) [8]).

Flexibility.   CNR (and its packet radio variant) has evolved to provide very flexible communications to combat forces in a wide variety of deployments and operational environments. All-informed communications can be provided across the deployed force enhancing the coordination of complex tasks. The additional range extension provided by airborne platform increases the usefulness of such systems. Satellite-based systems are constrained by their circuit-switched nature—the only thing a user can do with a satellite phone is make a phone call—this limits the usefulness of such systems in support of military operations as it is difficult (i.e., expensive) to form an all-informed voice net or a robust data network using circuit-switching.

Security.   All forms of mobile communications can be secured by the use of cryptographic equipment. Additional security is provided within the CNR nets by the fact that the users carry and own the equipment and that ranges are limited, reducing the opportunity of intercept. Satellite-based systems are very vulnerable to intercept and jamming and rely on infrastructure that is not owned by the tactical user, or even the service or defense force in most cases, which makes their availability very doubtful in times of crisis.

5.4   An Architectural Framework

The ideal tactical communications system architecture would provide a mobile infrastructure to support mobile users. It would therefore be a single homogeneous network supporting all communicating entities within the battlespace. As identified earlier, however, there are a number of difficulties in providing such an architecture.

A mobile infrastructure (CNR) provides ideal flexibility for voice and limited data communications between combat troops. However, this flexibility is gained at the cost of capacity and range. Therefore, CNR cannot provide the capacity required for useful data communications or for voice and data communications between command posts. In addition, ranges beyond those required for conventional deployments cannot be supported by CNR systems. The trunk communications system provides significant capacity to support the transfer of data between command posts. Trunk communications systems are not, however, sufficiently mobile to provide directly the intimate support required by combat troops.

While the tactical communications system can be provided as one logical network, it cannot be provided as one single physical network. At the lower level, combat troops carry a device that must be a network node as well as the access terminal. Battery power and the need for small omnidirectional antennas mean that ranges and capacities are limited. At the higher level, the large capacities required of trunk communications systems mean that they will remain semimobile for the foreseeable future. Large power requirements force the use of generators, and high-gain antennas must be deployed on guyed masts to provide reasonable ranges.

Neither the CNR subsystem nor the trunk communications subsystem is able to cover the large ranges required for dispersed or fast-moving operations. The only solution to providing high-capacity, long-range communications is to elevate the antennas. In the extreme, the provision of a satellite-based or an airborne repeater or switch will greatly increase the ranges between network nodes. A satellite-based solution is not considered desirable due to its inability to meet the requirements of a minimum organic communications system. An airborne subsystem is therefore required to support long-range operation. In addition, an airborne subsystem will increase the capacity of lower level tactical communications by removing the range restriction on high frequencies that can provide additional capacity from small omnidirectional antennas.

Current CNR nets and trunk communications systems do not provide an architecture that supports a large number of mobile voice and data users—the transfer of real-time situational awareness data is therefore correspondingly limited. The data-handling capacity of the trunk communications system will be sufficient (with some modification to the architecture) to cope with the volumes of data that must be transmitted between major headquarters. However, the CNR system's ability is severely limited, especially as it will still be required to transmit voice information. Therefore, an additional, purpose-designed, data distribution system is required to provide sufficient capacity to transfer situational awareness data across the lower levels of the battlefield. However, CNR must still be voice and data capable to allow organic communications of both types within subunits, should they be deployed individually or beyond the range of the tactical data distribution subsystem. The additional (albeit limited) data capacity in the combat radio subsystem would also provide an overflow capability, should the tactical data distribution subsystem be unable to meet all the data needs.

A minimum organic tactical communications system will be able to provide a basic level of service and must be able to be augmented where possible by overlaid communications systems such as the public telephone network, satellite-based communications systems, and PCSs. These overlaid systems cannot be guaranteed to be available and cannot therefore be included in the minimum organic system. However, if they are available, great advantage is to be gained from their use.

In order to simplify the user interface to these subsystems, a local communications subsystem (most probably containing a level of switching) is required. This local subsystem could take a number of forms, from a vehicle harness to a LAN around brigade headquarters.

To support command and control in future land warfare, the tactical communications system is therefore required to evolve from the two subsystems of Figure 5.5 to five subsystems [9]. The *combat radio subsystem* provides the mobile infrastructure to carry voice and data communications to support the command and control of combat troops. The *tactical data distribution subsystem* provides high-capacity data communications to support the situational awareness required for the command and control of combat troops. The *tactical trunk subsystem* provides the transportable infrastructure to support communications between command elements and other large-volume users. The *tactical airborne subsystem* extends communications ranges and provides additional capacity when the tactical situation allows. Finally, the *local subsystem* simplifies the user interface to the other communications subsystems and to overlaid communications systems.

The architecture of Figure 5.5 illustrates the major architectural components of the tactical communications system. It recognizes that, while the tactical communications system is to be considered as one logical network, for practical deployment reasons, it will be provided as a number of physical networks (at least in the short term). It is also a convenient starting point since it broadly coincides with the current deployed architecture, requiring the addition of a tactical data distribution subsystem as a high-capacity data system and a tactical airborne subsystem to increase capacity and range. However, the concept of a single logical network must remain paramount, as it is a crucial aspect of the architecture in Figure 5.5.

The tactical communications system does not exist in isolation; it exists to support a number of battlefield, joint, and combined systems. These supported systems interface to the tactical communications system as illustrated in Figure 5.5. As also illustrated in Figure 5.5, the minimum-essential

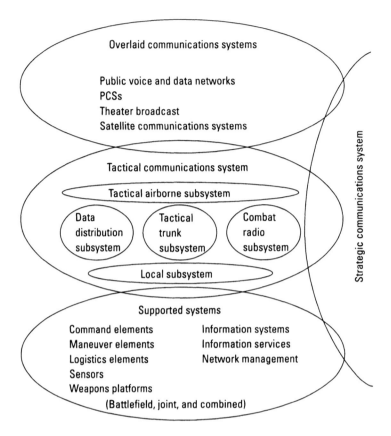

Figure 5.5 Elements of the tactical communications system and interfaces to other systems.

tactical communications system is augmented where possible by a range of overlaid communications systems such as the commercial telephone network, satellite communications, theater broadcast, and PCSs. These overlaid systems should be seamlessly integrated with the tactical communications system.

5.4.1 Supported Systems

The tactical communications system architecture must support a wide range of battlefield, joint, and combined systems.

5.4.1.1 Command, Maneuver, and Logistic Elements

Arguably, the principal purpose of the communications system architecture is to support the C2 cycle and the transfer of information between command, maneuver, and logistics elements. Interfaces between these elements and the tactical communications system are provided by the local subsystem, which is described in more detail in Chapter 10.

5.4.1.2 Sensors

In NCW, it is highly desirable for a user at any point in the battlespace to be able to access information provided by any sensor. However, this does not imply that raw sensor data is transferred across the tactical communications system, which simply cannot be provided with enough capacity to cope with large volume sensors connecting into the network at any point. What is required is a transfer of information, which does not necessarily require the transfer of the original sensor data. Each sensor system must be examined in the context of the architecture of the tactical communications system, with a view to determining the most effective way of interfacing that sensor to the network.

5.4.1.3 Weapons Platforms

Weapons platforms must be able to connect to any point in the tactical communications system and subsequently be able to access information from any sensor, the supported command element, and their own command post. Battlefield, joint, and combined weapons platforms exist in a variety of forms and in a variety of locations. Battlefield weapons platforms will be equipped with a local subsystem to allow access to a range of subsystems of the tactical communications system. The local subsystem will provide the range of interfaces required to allow the weapons platform to communicate to any other point in the battlespace, or to any point in the strategic communications system. Joint and combined weapons platforms must also be able to communicate to any point in the tactical communications system. For any particular army, further study is required to identify the range of weapons platforms, type of data required to be transferred, locations to which data/information is required to be transferred, and recommended types of links.

5.4.1.4 Information Systems

Information is provided to commanders by many different information systems, whether they are tactical, operational, or strategic systems. These systems extend from the tactical level to the joint level. In addition, access in the field is required to a wide range of administrative systems. The number of disparate systems is of great concern here. Before a more detailed architecture can be developed for any particular army, further study is required to address:

- Those information systems that intend to transfer data across the tactical communications system, either between systems within the tactical environment, or from outside the tactical communications system to an information system within;
- The nature of information transfer between systems;
- Suitable database architectures for tactical information systems;
- Limitations of the tactical communications system to support information transfer;
- Recommendations for the development of future information systems to be more suited to the transfer of information within the tactical environment.

5.4.1.5 Information Services

The tactical communications system must support the provision of vital information services throughout the battlespace. These services include security, messaging, video teleconferencing, data replication and warehousing, distributed computing, and search engines.

Security. The tactical communications system must have a security architecture that protects information sufficiently to allow information processing at a range of security levels, distributed processing, connectivity via unclassified networks, and interconnection of users whose security levels range from unclassified to the highest support classification. An appropriate security environment would allow users to exchange information in the safe knowledge that the data is authentic, originated from valid users, and is not available to those who do not have authorized access. The security architecture should define features that protect the confidentiality, integrity, and availability of information that is created, processed, stored, and communicated. In particular, the following major security services must be provided: data integrity, identification and authentication, nonrepudiation, data confidentiality, and access control.

Messaging. The success of modern command systems depends on the efficient transfer of information. In an ideal world, all communicating entities would have the same databases, operating systems, and computing platforms. This is rarely the case within the same service, however, let alone within a joint or combined environment. Interoperability between disparate systems can be assured through the use of messages, which are the primary means of transferring information between battlefield locations. Message text formats (MTF) are defined by a number of standards, including the NATO standard Allied Data Publications Number 3 (AdatP-3), Australian Defense Formatted Message (ADFORM), and U.S. Message Text Formats (USMTF) standards [10]. MTFs have a structured format based on a well-defined set of rules and every message type is constructed from the basic set of rules. There are several hundred message formats within each standard including situation reports, intelligence reports, and air tasking orders. Importantly, for interoperability between systems with different levels of automation, MTFs have been designed to be readable by both computers and human recipients. Although AdatP-3 and ADFORM standards were originally derived from USMTF, there are a number of differences between the standards and considerable time and effort are spent by the United States, the Australian Defense Force (ADF), and NATO to ensure the consistency and availability of the MTF standards. In many armies there is often little interoperability between information systems due in part to the use of proprietary communications formats. This situation must be addressed. When it is, the question of an appropriate message format arises. The use of bit-oriented messaging is essential within the tactical communications system if near-real-time situational awareness is to be feasible. Further study of this important area is required to address suitable message formats for use within the tactical communications system, techniques for retaining interoperability with command systems at theater level and above (including systems within the strategic communications system), and joint and combined systems [particularly in light of U.S. use of tactical data link (TADIL-J) series of standards].

Videoconferencing.   Battlefield videoconferencing is a service that is considered by the modern commander to be very important, if not essential. Through video teleconferencing, commanders can effectively disseminate orders and can conduct collaborative planning and white-boarding with subordinate commanders and key staff elements. However, there have been few cost-benefit analyses to support the introduction of such a bandwidth-hungry application into the tactical environment. It is unlikely that there will ever be sufficient bandwidth available within the combat radio subsystem to support videoconferencing. Given that commanders at battalion level have only mobile access to the tactical trunk subsystem, videoconferencing from brigade to battalion will not be possible without the provision of a dedicated system. Sufficient bandwidth should be available for videoconferencing for fixed subscribers to the tactical trunk subsystem. Further study is required into the cost benefits of videoconferencing in the tactical environment through the tactical communications system. The following issues need to be addressed:

- The operational requirement for battlefield videoconferencing;
- The ability of the tactical trunk subsystem to carry videoconferencing between fixed subscribers;
- The ability of the tactical trunk subsystem to carry videoconferencing between mobile subscribers;
- The requirement for a dedicated videoconferencing system, perhaps utilizing the tactical airborne subsystem.

5.4.1.6 Network Management

As trunk networks have matured and become less aligned with the chain of command, network management has also had to evolve. Indeed, the management task has become more complicated as the networks have evolved. The move away from the chain-of-command network has meant that network management must be conducted separately, and, in the case of area networks, it must also be conducted centrally. The underlying theme of the development of the tactical communications system is the provision of a seamless network of networks that provides rapid and efficient transfer of information between any two points in the battlespace and between any point in the battlespace and any point in the strategic network. Since the terminals and the nodes of the network move at very frequent intervals, it is essential that the network is kept in good working order and that it does not fragment, or that bottlenecks do not appear. This requires a network management infrastructure that can plan, install, monitor, maintain, and troubleshoot the network to support a wide spectrum of operations.

5.4.2 Overlaid Communications Systems

The tactical communications system provides the minimum essential communications for a deployed force, using equipment and personnel that are organic to that force. Additional communications capacity can often be obtained from a variety of systems that are either of commercial rather than military design or are provided by external sources. In this book, these additional communications resources are referred to as *overlaid communications systems.*

Overlaid systems augment tactical capability as rapidly as possible. For example, during the Vietnam conflict, nearly every U.S. division-sized force depended on external communications support to augment internal resources, particularly to support base camp areas while integral assets were concentrated on operational requirements [11]. Additionally, separate brigade-sized units were frequently deployed with minimal organic communications support, which needed augmentation if they were given any extensive combat missions. During the Gulf War, tactical communications systems were quickly integrated into strategic systems (including two commercial satellite terminals and 15 microwave systems) and corps-level systems including the Defense Communications System (DCS), long-range single-channel radios, and tactical satellite assets [12]. In particular, the number of satellite systems employed was far beyond expectations and satellite communications were judged the ideal distance and terrain-independent communications solution [13].

An overlaid communications system is a communications system that is not organic to the deployed tactical force, generally designed and manufactured to meet commercial rather than military needs, or managed, operated, and maintained by an external organization rather than directly by the defense force by which it is being used. There are a number of key differences between the tactical communications system and overlaid communications systems.

Electronic protection (EP).   EP techniques, such as frequency hopping, are normally considered to be an essential component of the design of modern military communications systems. Commercial systems, however, are not designed to operate in a hostile electromagnetic environment. Overlaid communications systems involving wireless links are therefore vulnerable to adversary EW. EP features offered by typical military communications equipment include frequency hopping and DSSS. Overlaid communications systems may sometimes appear to offer similar features. One example is the use of DSSS to provide multiple access in a CDMA mobile telephone system. A crucial difference, however, from a military implementation of DSSS is that the spreading sequence in a CDMA mobile telephone system is much shorter than would be found in a military system and is provided as a multiple-access technique rather than as an EP measure. Commercial spread-spectrum systems are also not secure and therefore do not have any advantages over other commercial systems in terms of EP. In some circumstances, such as a military communications satellite, the overlaid communications system may be specifically designed for military applications, and may incorporate EP. The lack of EP in some overlaid communications systems leaves them vulnerable to jamming and other effects of electromagnetic radiation and may restrict their use in hostile electromagnetic environments. This applies to all systems, not just those involving wireless connections.

Control. An overlaid communications system is most likely to be operated, managed, and maintained by an organization external to a deployed force. This may be a commercial organization, often based in a foreign country, which raises a number of issues related to the reliability of service that will be offered by the overlaid communications system. These issues cannot be overcome through confidentiality measures such as encryption. In some circumstances a force may provide its own overlaid communications system, which may occur where the force deploys commercial equipment to supplement the tactical communications system. This may be an attractive option where there is a low electromagnetic threat due to the lower cost of these commercial systems compared to military-specific systems of the same capacity.

Security. Current commercial communications equipment does not incorporate military-grade encryption. Additionally, some electronic surveillance threats are difficult to counter when communications pass via commercial switching centers. Secure communications over an overlaid communications system may be achieved by provision of bulk encryption at the interface between the overlaid communications system and the tactical communications system, or procurement of military terminals that incorporate encryption for direct access to the overlaid communications system.

Ruggedization. Because of its commercial origin, equipment in an overlaid communications system is typically not ruggedized to military standards. It is therefore more vulnerable to effects of immersion in water and extremes of temperature, vibration, shock, pressure, and humidity. This major disadvantage may be overcome in some circumstances by the procurement of ruggedized systems capable of accessing overlaid communications systems. This may be achieved by provision of an interface between the overlaid communications system and the tactical communications system that enables user terminals attached to the tactical communications system to access the services of the overlaid communications system, or provision of special ruggedized terminals that access the overlaid communications system directly (e.g., a ruggedized GSM handset).

User interface. Commercial user terminals are typically not operable with nuclear, biological, and chemical (NBC) gloves/hood. This is exemplified by commercial cellular telephones, whose keypads are now so small as to be difficult to use for those with large fingers. This difficulty can be overcome by the procurement of military terminals for direct access of the overlaid communications system (e.g., a ruggedized GSM handset), or the provision of an interface between the tactical communications system and the overlaid communications system that enables user terminals attached to the tactical communications system to access the services of the overlaid communications system.

Physical security of infrastructure.   Much of the infrastructure of an overlaid communications system will often be outside the area that is secured by a deployed force. In some circumstances, such as a microwave retransmission site, it may be possible for the deployed force to provide protection for this infrastructure. In others, such as optical fiber cables running over long distances, this may not be possible.

5.5 Summary

This chapter has discussed the key design drivers for the development of an architecture for the tactical communications system. These design drivers include the traditional principles of military communications as well as a number of important issues governing the way in which the tactical communications system is to be employed.

The tactical communications system must be organic to the supported force and must support communications between any two points in the battlespace and between any point in the battlespace and the strategic communications system. Communications support must be provided to a range of battlefield, joint, and combined systems. Access must also be gained to a range of additional overlaid communications systems to increase the capacity of the minimum organic network.

While it is essential that the tactical communications system provide a single logical network, it is not possible to provide a single physical network. The range of candidate technologies available to provide access to mobile users constrains the physical architecture to the provision of two major subsystems—the tactical trunk subsystem and the combat radio subsystem. To extend the range of these two subsystems in dispersed operations, a tactical airborne subsystem is required. Additionally, there is not sufficient capacity, in the combat radio subsystem in particular, to cope with the high volume of data transfer required to support real-time situational awareness for commanders of combat forces—this need is met by the tactical data distribution subsystem. The local subsystem simplifies the user interface to the other communications subsystems and the overlaid communications systems. Each of the subsystems of the tactical communications system is considered in more detail in Chapters 6 through 10.

Endnotes

[1]   Here we define the upper-level view of the systems architecture in the context of the C4ISR Architectures Working Group, *C4ISR Architecture Framework*, Version 2.0, Washington, D.C.: U.S. Department of Defense, Dec. 18, 1997.

[2]   "Joint Spectrum Center Eases Foxhole Frequency Allocation," *Signal*, Vol. 51, No. 6, 1997, p. 59.

[3]   For example, see U.S. use of HF in Vietnam War as described in Bergen, J., *Military Communications, A Test for Technology*, Washington, D.C.: Center for Military History, United States Army, 1986, p. 280.

[4]   ANSI/IEEE Std 802.11, "Wireless LAN Medium Access Control (MAC) and Physical Layer (PHY) Specifications," Piscataway, NJ: IEEE, 1999.

[5]   "Specification of the Bluetooth System," Bluetooth SIG, Version 1.1, Feb. 2001.

[6]   The classifications provided here are sufficient for our discussion of a suitable architecture. A more detailed analysis in a similar vein can be found in Feldman, P., *Emerging Commercial Mobile Wireless Technology and Standards: Suitable for the Army?* RAND Corporation Report MR-960-A, 1998.

[7]   Third-generation mobile communications systems promise up to 2 Mbps in the smallest cells (picocells).

[8]   Edwards, T., "More Than Just Hot Air," *Communications International*, Sept. 1999, pp. 42–45.

[9]   Ryan, M., and M. Frater, "An Architectural Framework for Modern Tactical Communications Systems," *IEEE Military Communications Conference (MILCOM 2000)*, Los Angeles, CA, Oct. 23–25, 2000.

[10]   For a good overview of message formats, see: Peach, W., "Message Text Formats—A Solution to the Problem of Interoperability," *Journal of Battlefield Technology*, Vol. 2, No. 1, March 1999, pp. 11–16.

[11]   Meyer, C., *Division-Level Communications 1962–1973*, Department of the Army: Washington, D.C., 1982, pp. 27–28.

[12] *The U.S. Army Signal Corps in Operation Desert Shield/Desert Storm,* U.S. Army Signal Center Historical Monograph Series, Office of the Command Historian, United States Army Signal Center and Fort Gordon: Fort Gordon, GA, 1994, pp. 5–6.

[13] *The U.S. Army Signal Corps in Operation Desert Shield/Desert Storm,* U.S. Army Signal Center Historical Monograph Series, Office of the Command Historian, United States Army Signal Center and Fort Gordon: Fort Gordon, GA, 1994, p.14.

6

Tactical Trunk Subsystem

6.1 Introduction

Traditionally, the tactical trunk subsystem is the principal means of communication down to formation/unit and logistic installation level. The subsystem comprises multichannel radio equipment, line, switches, and terminating facilities to provide voice, telegraph, facsimile, video, and data communications as well as hand carriage.

The tactical trunk subsystem provides high-capacity communications to a wide range of types of communications. Traditionally, the successful negotiation of the C2 cycle has relied on voice communications that allow commanders and staffs to speak to each other. Networks have also facilitated the exchange of documents and text via facsimile and telegraph messages. While interpersonal voice communications are still essential, facsimile and telegraph are increasingly being replaced by the exchange of data between information systems. These intersystem communications are becoming more important as computing power increases and more computers and other data-oriented systems such as sensors are deployed on the battlefield.

The position of the tactical trunk subsystem in the range/capacity/mobility trade-off is shown in Figure 6.1. Reliable, high-capacity links require the use of high frequencies with high-gain antennas. Longer ranges also require the use of higher powers than can be obtained from batteries and some form of generator-based power supply is necessary in the communications detachment. Since tactical communications are more often terrain-

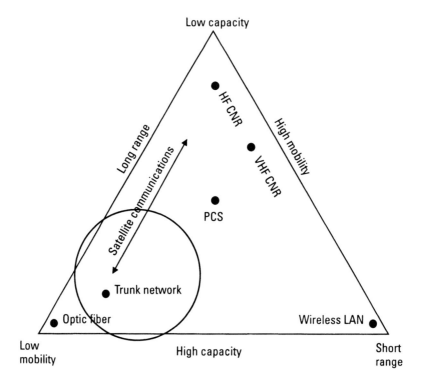

Figure 6.1 Position of the tactical trunk subsystem in the range/capacity/mobility trade-off.

limited than it is power-limited, long-range communications also requires the use of mast-mounted antennas. In all, therefore, trunk communications systems must have mast-mounted antennas and generator-based power supplies, which means that they are semimobile at best since mobility is traded off to achieve capacity and range.

Despite rapid advances in technology, there are still many battlefield communications that are too large or not in a suitable form to be transmitted over radio or line links. These communications are transferred between headquarters by the SDS. While SDS remains an important battlefield communications means that will most probably always exist on any future battlefield, it is not discussed further in this book.

This chapter provides a brief examination of the design drivers from Chapter 5 to determine their impact on the tactical trunk subsystem. Discussion then centers on architectural options to provide a tactical trunk subsystem that is able to meet the spectrum of operations.

6.2 Key Architectural Drivers

Chapter 5 described the design drivers of the tactical communications system. This section considers the impact of those key drivers on the tactical trunk subsystem.

6.2.1 Communications Support the Chain of Command

As discussed in Chapter 4, modern area trunk networks have evolved to provide a topology of meshed nodes to ensure that communications support the chain of command. The trunk network deploys to cover the area of operation, and headquarters can move about within the coverage of the network to meet the demands of the tactical situation, connecting by radio relay to the nearest node. Commanders can therefore deploy as required by the tactical situation and are almost completely unfettered by their communications. Additionally, command relationships can be changed quickly without needing to redeploy communications elements between headquarters and units under command. These capabilities become even more important on the modern battlefield. As discussed in Section 6.3, a key concept within modern trunk systems is the separation of the network's trunk-interconnection function and the communications-access function that is integral to the headquarters and is logically (if not physically) separate.

6.2.2 Integration

The tactical communications system must operate as an integrated network offering seamless connectivity between battlefield elements as well as between battlefield elements and the strategic communications system. The requirement for integration has a number of impacts for the design of the tactical trunk subsystem.

Interfaces. The tactical trunk subsystem will take the major responsibility for the provision of seamless connectivity within the tactical communications system. In general, the terminals in the combat radio subsystem will be too small to be able to incorporate additional components for interfaces. Similarly, the interfaces required for the tactical data distribution subsystem, the overlaid communications system, and the strategic communications system will also need to be contained within the tactical trunk subsystem. Detailed discussion on the interfaces required between subsystems and systems is contained in Chapter 10.

Support for real-time and non-real-time communications. The tactical communications system is required to support both real-time communications (e.g., voice and video) along with near-real-time or non-real-time communications (e.g., messaging and file transfer). Consequently, the trunk system is required to cater to the varying needs of each of these types of services. This will particularly impact the requirements for switching and the types of services provided.

Network management. One of the most critical elements required for seamless integration is an appropriate network management system. Communications must therefore be controlled at the highest level of deployed force, through an integrated network management structure that supports the chain of command. Complementary tactical and technical control must be exercised throughout the whole communications system. As a central component of the overall network, the tactical trunk subsystem must play a key role in the provision of network management facilities. This crucial area is very often overlooked in the design of a system and is often not fielded until well after the trunk network enters service.

6.2.3 Reliability

While all communications hardware must be reliable, reliability for a tactical trunk network is more driven by the network topology. In fourth-generation networks the mesh topology significantly reduces the potential for networks to be vulnerable to single points of failure and to allow individual network components to move as required by the tactical situation. Further reliability is provided by the provision of redundancy in the switching nodes, and through the use of alternative transmission means for the bearer services.

6.2.4 Simplicity

While the elements of the trunk network do not need the mobility of those in the combat radio subsystem, they must be able to deploy and redeploy rapidly. Therefore, the various discrete components of the network (i.e., bearers, switches, and node control centers) must be designed to allow the network to adjust to fluid changes in tactical situation. While the functions required of modern trunk networks become more complex, their architecture must remain easy to deploy and operate and simple to repair.

6.2.5 Capacity

The trunk system provides the interconnection between headquarters staff at various levels and the capacity required to serve each headquarters is a key design driver of the trunk subsystem, particularly in light of the increasing deployment of modern command-support and high-volume sensor systems. Traditionally, only headquarters from brigade and above had sufficient demand for capacity to warrant interconnection by an access node to the trunk network. Indeed, the lowest level of trunk subscriber has normally been the battalion commander through SCRA. In recent years, however, the deployment of large numbers of information systems and sensors has generally meant that the demands for wide-area connection between lower-level headquarters are well in excess of the capability of the combat radio subsystem. The tactical trunk subsystem must therefore cope with higher capacities to more headquarters, which increases the overall capacity of the network. The need for high capacity has the following impacts on the tactical trunk subsystem.

Switching. Network switches must be able to cope with large volumes of traffic with both real-time and non-real-time requirements. The backbone switching architecture must also be able to be scaled easily to cope with wide variations in network capacity as well as types of deployment.

Bearers. The higher capacities required between nodes force the use of bearers in higher frequency bands with higher signal-to-noise ratios. In turn, these requirements force the use of higher-gain antennas that are more difficult to deploy and orientate. In all, these factors result in the reduction of the range of terrestrial trunk radio bearers, at a time when networks are required to cover wider areas of operation. Long-range bearers are difficult to provide for a number of reasons:

- HF radio provides long ranges but with bandwidths that are far too low to be useable in trunk networks, except perhaps for last-ditch communications.

- Troposcatter systems can provide high-capacity long-range communications (up to 200 km for tactical systems). The terminals are large, however, and require large parabolic antennas that are not very mobile.

- Longer ranges are possible using satellite communications, although there is only a limited capacity available from these systems (limited mostly by the platform's ability to carry the transponder). Therefore, in the tactical trunk subsystem, the use of terrestrial systems must be maximized, reserving satellites for those links for which terrestrial communications are not possible.

- Range can also be extended by airborne repeater systems, although capacity also tends to be limited by the platform's ability to carry and power repeaters.

Bandwidth management.   Since capacity will always be limited by the practical provision of the network, there is a need for intelligent management of this resource. Staff officers and network managers must be equipped with appropriate automated tools to allow for bandwidth management.

6.2.6 QoS

The radio bearers of the trunk network carry large volumes of data communications. Whereas voice communications is generally very robust and effective communications can be maintained over quite noisy links, data communications requires the addition of significant levels of overhead to provide the error protection required to accommodate noisy channels. This error protection overhead consumes bearer bandwidth and should be as low as possible and only applied when necessary. Additionally, some services impose tight controls on transmission delay, which must be supported by the trunk network, yet other services can cope with quite significant delays. The selection of appropriate data communications protocols is therefore particularly important and the requirement of providing varying QoS requirements is perhaps the most significant design driver in modern networks.

6.2.7 Mobility

The support for mobility could philosophically be provided in one of two ways.

1. The users' mobility could be supported by providing a ubiquitous network that surrounds all the users and is available to terminate any radio connection as required. This is the approach taken with commercial PCS networks, where a large fixed infrastructure is always available to the users who are free to roam within it.

2. The network can be given sufficient mobility to be able to keep up with the mobile users and provide a network that supports user

connectivity within the area of operations. This latter approach is the one taken by military trunk communications networks as the most cost-effective way of serving a combat force across a wide spectrum of operations.

The tactical trunk subsystem must therefore be as mobile as the forces it supports. The components of the subsystem must therefore not only be flexible, modular, and easily reconfigured, but they must be able to support forces with varying degrees of mobility—from foot-mounted, to motorized to mechanized/armored formations.

6.2.8 Security

A key concern is that the security architecture in the tactical domain may be driven by different imperatives than those in the strategic domain. Care must be taken in developing the trunk system technical architecture to ensure that interoperability with other communications and information systems (battlefield, joint, and combined) is not prejudiced by incompatible security architectures.

6.2.9 Survivability

Since command and control on the modern battlefield are heavily reliant on electronic communications, the survivability of the tactical trunk subsystem is paramount if interheadquarter communications are to be maintained. Provision of a robust survivable network traditionally relies on a significant level of redundancy in network components to ensure that loss of a number of nodes does not significantly affect network performance. It would be reasonable to expect a trunk network to be designed to continue to provide a 100% grade of service with only 50% of its nodes available.

6.2.10 Communications Support for the Spectrum of Operations

Current network topologies were developed to suit conventional high-intensity deployments, notably in northwest Europe during the Cold War. Within these environments, a corps network would be expected to have some 40 trunk nodes to guarantee coverage of fast-moving forces within an area of operations that could be as small as 50 km by 70 km. In dispersed operations, however, it would not normally be feasible to provide the significant numbers of trunk node assets across the entire area of operations to allow commanders to redeploy without significantly affecting the network structure. As a measure of comparison, nearly 1,000 trunk nodes would be required to cover an area of operations with a 500-km radius. For example, when the British Army deployed to the Gulf War, although only a reduced-size division of two brigade groups were deployed, the distances involved were so great that it required almost all of 1(BR) Corps' communications assets to meet the requirement. Even so, Ptarmigan trunk nodes were so overstretched that they could not have provided a network without the deployment of satellite bridges to link nodes together [1].

Trunk networks will therefore have to be provided with more sparsely populated topologies in dispersed operations, and supported headquarters and users may be somewhat restricted in their ability to move compared to high-density operations. Nevertheless, it is still important to discriminate between the communications-access function, which is integral to the head-quarters, and the trunk-interconnection function, which is logically (if not physically) separate.

The tactical trunk subsystem must continue to be able to support high-intensity operations, which will call for quite short ranges and a densely populated trunk network, as well as be able to deploy with minimal changes to support widely dispersed operations, which will tax the subsystems to pro-vide a resilient network topology. While remaining cognizant of the need to minimize the number of communications vehicles in headquarters, the trunk system architecture therefore needs to be modular to provide the flexibility to deploy in different configurations.

6.2.11 Command and Control on the Move

There are three distinct needs for mobile communications on the battle-field. The first, directly between combat commanders, is served by the highly-mobile combat radio subsystem. The second need is to provide com-munications between this level of command and higher levels, which is served either by extending the combat radio subsystem or by the intercon-nection of combat radio nets via the trunk subsystem. The third require-ment is between a mobile user and a fixed trunk user, either by allowing the use of a combat radio as the mobile terminal operating through a gateway (currently CNRI) or as a direct trunk interface using technology such as SCRA or the commercial PCSs.

The major difficulty with supporting command and control on the move is the ability to provide continuous network access to a mobile com-mander. In commercial networks, users are supported by the provision of a massive amount of fixed infrastructure within which the user may move at will without losing the ability to communicate. The military equivalent is much harder to engineer due to the inability to provide fixed infrastructure, and the lack of coverage of the forward regions of the battlefield that is forced by the inability to deploy infrastructure within the adversary's portion of the area of operations.

6.2.12 Size of Supported Force/Organic, Minimum-Essential Communications

Traditionally, trunk communications are provided as a corps-level network, covering the corps area. Of the 40 or so trunk nodes deployed, each brigade would deploy with one or two. However, the relationship between brigade headquarters and trunk node is normally only for the practical deployment of the network. The technical control of the node is retained by the corps signals organization. For many modern deployments, the trunk network must be viewed from a different perspective—the tactical trunk subsystem is required to provide organic, minimum-essential communications for the supported force, whether it is of brigade, division, or corps size. This requires the concept of a division-level and brigade-level trunk network—a concept that is only beginning to emerge in most modern armies.

6.3 Architectural Overview

6.3.1 Network Entities

A tactical trunk subsystem comprises four main entities: *access nodes* to provide local switching for a community of users and to connect that community to *trunk nodes,* which provide backbone switches that are interconnected by *bearers* to interconnect nodes into a network. Finally, *RAPs* provide single-channel radio access to mobile trunk subscribers. (Note that the term *switch* is used here as a generic label for a device that enables interconnection between users/user applications—possible switching architectures are discussed in Section 6.6.) In principle, a user or group of users should be able to connect to the communications system at any point on the battlefield and have seamless access to any other user or group of users who are similarly connected without needing to know precisely where any other user might be located. Additionally, while the network of trunk nodes may need to reconfigure to accommodate the tactical plan, the movement of users or switching nodes should not impact on the performance of the network.

The separation of the user community and the switch network is often depicted as a cloud (see Figure 6.2) to emphasize that the user community is not necessarily aware of the network topology or of the location of nodes within the cloud. Fourth-generation trunk networks separate users from the communications backbone through the use of access nodes and trunk nodes. The access node provides local switching to its community of users and then aggregates their use to present this community to the trunk network. The trunk nodes are then meshed to provide the communications infrastructure through which the access nodes (user communities) are connected. In such a system, the physical separation of access nodes and trunk nodes reflects the logical separation of user and communications infrastructure.

In most major armies brigade headquarters is the lowest level to which access nodes have been provided. Access at lower levels is normally only available through the mobile subscriber facilities provided by SCRA. For most modern deployments, however, the trunk network must extend to lower levels to increase the capacity available between lower-level headquarters. So in addition to serving corps- and divisional-level headquarters, modern trunk networks must support major communications nodes (access nodes) below brigade level at battalion headquarters, logistics installations, and headquarters of other major units in brigade areas of operations. Minor communications nodes may also be required at company headquarters and below, although sufficient service can generally be provided at this level by mobile access.

There are a number of architectural options for the provision of a tactical trunk communications network. In particular, there are design options for types of bearers, nodal topology, node composition, and mobile user access.

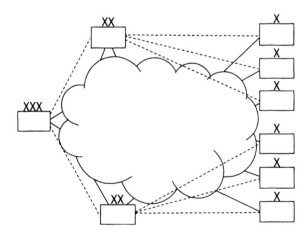

Figure 6.2 A fourth-generation trunk network represented as a cloud.

6.4 Types of Bearers

In consideration of suitable architectures for tactical trunk networks, bearers are considered first because the limitations of various bearers will affect the other architectural options. To provide wide area communications, there are a number of bearer options: HF radio, fiber-optic cable, VHF/UHF radio relay, tropospheric scatter, and satellite. Several of these systems can have their communications ranges extended by the use of the tactical airborne subsystem (as discussed in more detail in Chapter 9).

6.4.1 Military Utility

Each of these bearers has different military utility.

Capacity.   High data rates are required between trunk nodes. Based on experience with most trunk networks, data rates of at least 2 Mbps are required to support communications between trunk nodes. Future requirements are likely to increase the necessary data rates by an order of magnitude. HF radio does not have sufficient capacity to support communication between trunk nodes, and is really only useful for last-ditch communications at very low data rates. Fiber-optic systems are able to provide many times more than this capacity. VHF/UHF radio relay and tropospheric scatter systems have sufficient capacity to support these rates. Satellite communications can also support these rates, although, as a system, satellite capacity is constrained by the capability of the transponder and a number of technical parameters. It does not all follow, therefore, that a large number of 2-Mbps links could be supported by a single satellite, particularly when the ground terminals are widely dispersed.

Reliability.   Apart from HF communications, which is generally too unreliable for use as a primary trunk bearer, all types of bearers have high reliability.

Quality.   HF system link performance will impact on quality, as bandwidth constraints will force the use of lower-quality voice-coding schemes. All other bearers can be engineered to provide acceptable quality.

Communications support for the spectrum of operations.   Deployment within a wide spectrum of operations will call for a variety of operating ranges from communications bearers. Short ranges are relatively easy to obtain with all types of bearers, with radio relay representing the best compromise between system complexity and capacity. The longer ranges required to support dispersed operations, however, cannot be totally supported using line-of-sight bearers (even using two or three hops). For longer ranges, HF or satellite communications are required.

Command and control on the move.   Clearly, fiber-optic cables are totally unsuitable for mobile communications. Each of the other systems can support mobile operations—but only by moving between bounds—and they typically cannot operate in a high capacity mode when totally mobile. In that regard, most bearer systems are generally considered to be semimobile. The exceptions are with HF and low-capacity (small antenna size) satellite applications, as well as the SCRA interface to trunk networks.

Organic, minimum-essential communications.   Satellite systems inevitably depend on the satellite transponder, which cannot be considered to be organic to the deployed force. For this reason, satellite communications are not considered suitable for inclusion in the tactical communications system, but are included in the overlaid communications system. All of the other bearers are able to be included as part of organic systems.

Jamming resistance.   Jamming is not an issue for optical fiber systems due to the difficulty of the jammer in gaining access to the channel. HF communications are also relatively difficult to jam due to the powers required by the jammer and the management systems that are employed by the communications systems to accommodate variations in the ionosphere. VHF/UHF radio relay bearers are possibly easier to jam because the jammer can be deployed closer. However, the use of high-gain antennas in these systems allows them to be deployed so that they can largely ignore potential jammers [2]. Troposcatter communications are inherently resistant to jamming due the propagation mechanism and their narrow beamwidth. Satellite systems are inherently much more vulnerable to jamming than any of the other systems, largely because a jammer can be located anywhere in the footprint of the transponder. The provision of jamming resistance on satellite links is possible, but significantly reduces the bandwidth available. In particular, most commercial FDMA satellites will provide little or no protection against jamming.

Economy.   In the tactical environment, optical-fiber communications are too expensive to deploy and recover over any distances longer than 1 km, which means that fiber optics is of little use in wide-area networking for mobile communications. HF communications are relatively very cheap, but lack bandwidth. The relative order of costs for the other systems will depend on the architecture chosen and the extent to which the network can expand to cover large areas. Traditional radio relay is cost-effective when ranges are relatively short. In widely dispersed operations, however, too many trunk nodes are required. Troposcatter links provide longer ranges and fewer nodes, although the cost of owning and operating each node is considerably higher than radio relay. Satellite communications are much better suited for providing high-capacity communications over long ranges, but it is the most expensive of the systems when the costs of the satellite are included in the system costs. Relay through the tactical airborne subsystem also has the potential to be expensive when the airborne platform is included. No single bearer system remains cost-effective while meeting all deployment scenarios. For example, for conventional operations, radio relay is most suitable. For widely dispersed operations, range extension is required through the use of troposcatter, the satellite overlaid system, or the tactical airborne subsystem.

6.4.2 Preferred Option

HF communications are unsuitable because of the low bandwidth and poor quality, although a role may be possible if a last-ditch backup communications system is required, particularly over long ranges. Optical fibers are unsuited to wide-area networking and are only able to be used in short runs (less than 1 km).

Satellite communications are not suitable for inclusion as part of the organic minimum-essential communications system because the satellite cannot be considered organic to the deployed force, even in cases where the interface (i.e., the satellite terminal) is organic. However, the use of satellite communications in the overlaid communications system is very useful to extend ranges of internodal links when required in dispersed operations.

Troposcatter communications do not have the setup and teardown times required to support the deployment of trunk nodes. However, troposcatter communications can provide high bandwidths (up to 2 Mbps) over long ranges (150–200 km for tactical systems). The utility of troposcatter is such that it would be a very useful adjunct to the organic assets of the tactical communications system. While troposcatter systems do not necessarily have to be provided on a scale to cover links between all trunk nodes, a small number of troposcatter terminals should be able to be provided within each deployed force so that at least one troposcatter link can be established as well as an additional terminal to anchor a link to a higher formation.

The most suitable bearer for the organic minimum-essential trunk communications is VHF/UHF terrestrial radio relay. Even so, the range of these systems is limited to line-of-sight and needs to be extended for dispersed operations by the organic tactical airborne subsystem, organic troposcatter communications, or satellite communications (as an overlaid communications system).

6.5 Nodal Topology

The tactical trunk subsystem is required to provide a high-capacity trunk communications network to support access nodes at brigade headquarters, battalion headquarters, major logistics installations, and headquarters of other major units within the brigade area of operations. Mobile access is required for battalion headquarters and below. There are two broad options for the provision of such a network: meshed or hub-spoke.

- *Meshed.* As illustrated in Figure 6.3, in a meshed topology the nodes are connected together such that each node is connected to at least three other nodes to provide a number of alternate routes through the network. The number of nodes in the network and degree of meshing dictates how many internodal links are required between any two points of the network, as well as the number of alternate paths that could provide the same end-to-end connection. The larger the number of nodes, the more flexible and survivable the network becomes.
- *Hub-spoke (or star).* As illustrated in Figure 6.4, the hub-spoke topology has a number of spoke nodes, each of which is connected to a common hub. The topology requires only that each node can communicate with the hub. There are only ever two internodal links between any two users in the network.

6.5.1 Military Utility

Capacity.   The capacity of each type of topology is similar, although the switching nodes in each network would be optimized differently. The major constraint on capacity would depend on the type of bearer chosen to interconnect each node. For radio-relay bearers, the meshed network can provide the greatest capacity because of the redundancy in bearers and switching. For

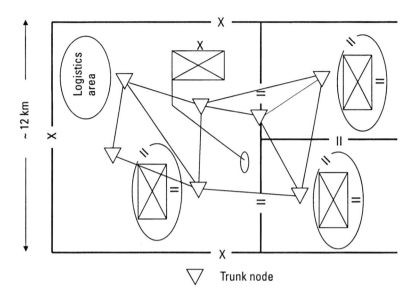

Figure 6.3 An example of a meshed network in a conventional brigade area.

satellite-based bearers, however, the hub-spoke configuration can provide greater capacity from a given satellite if small mobile satellite terminals are connected to a large dish station at the hub, rather than to another small dish as they would in a meshed network. If both terrestrial and satellite bearers are required, concurrent use of terrestrial mesh with hub-spoke elements would clearly provide maximum network capacity.

Reliability. A meshed network provides the highest reliability due to the existence of a number of alternative routes between any two points in the network. This advantage tends to diminish, however, as the network is strained, either by losses of nodes or by rapid movement that may prohibit the complete meshed network from being available continuously. The hub-spoke is generally less reliable due to a potential single point of failure, although this can be addressed by the provision of alternative hubs. If the hub is located out of the area of operations (perhaps within the strategic domain), this vulnerability is further reduced and can be addressed mostly by equipment redundancy. The hub-spoke configuration may also be more reliable in widely dispersed or fast-moving deployments where the meshed network planners struggle to maintain the interconnectivity between nodes.

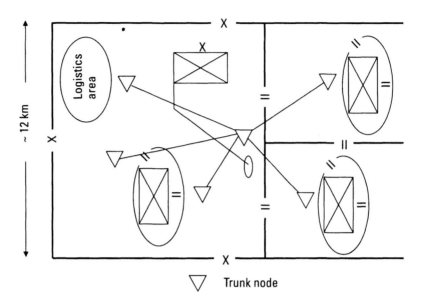

∇   Trunk node

Figure 6.4 An example of a hub-spoke network in a conventional brigade area.

Flexibility.   Fourth-generation trunk networks have evolved as meshed networks because they are extremely flexible and can accommodate the movement, or destruction, of any one node without significant disturbance of the network. In a well-planned network, a number of nodes can be moving without disrupting service. However, the hub-spoke topology constrains the flexibility of a network based on radio-relay bearers because the hub cannot move without disrupting the entire network. Hub-spoke topologies are therefore not favored for use with organic terrestrial bearers. For satellite bearers, however, the situation tends to be reversed as the hub-spoke topology does not suffer at all when one node moves or is destroyed. Additionally, the topology can cope with movement or destruction of a satellite-based hub because transferring communications from one hub to another redundant hub within the satellite footprint is straightforward. If a strategic hub is employed, then no movement of the hub is required. In addition, meshed topologies are not preferred for satellite bearers because the uplink and downlink delay normally limits internodal links to two hops, which creates a sparsely meshed network that has few of the advantages of such topologies.

Integration.   The two main requirements for integration for the trunk subsystem are the internal aspects of integration with the combat-radio tactical data distribution and tactical airborne subsystems, and the external aspects of integration with overlaid and strategic communications subsystems. For meshed networks, these issues generally relate to the location of suitable interfaces within the network to these other subsystems and systems. The advantage of a meshed network is that the mesh provides a robust way to interconnect any point of the network with the appropriate interface, no matter where that interface is located. Within the hub-spoke topology, interfaces are naturally best supported at the hub, which can increase the difficulties with flexibility, reliability, and survivability. These difficulties may be eased if the hub is located in the strategic domain, where the support of interfaces is easier, particularly in terms of integration between the tactical communications system and overlaid and strategic systems. If the hub is located in the strategic environment, however, internal interfaces with the combat radio, tactical data distribution, and tactical airborne subsystems will be difficult to arrange, as these will have to be supported at each of the spoke nodes.

Communications support for the spectrum of operations.   The meshed topology based on radio-relay bearers is the most suitable for high-density operations and almost all modern trunk networks have evolved in that form. Radio-relay bearers are not suitable for all modern deployments, however. For dispersed operations, range extension must be provided by organic assets (tactical airborne subsystem and troposcatter) or overlaid communications system assets such as satellite links.

Organic, minimum-essential communications.   Organic bearers include radio relay and troposcatter. The meshed topology is the best topology for these bearers, for the reasons outlined above. However, in some circumstances (particularly when terrain is not controlled in dispersed operations), a hub-spoke topology may be the most suitable. Of course, the hub-spoke topology is the most suitable when satellite bearers are used for range extension.

Survivability.   Meshed networks are very survivable as each node is connected to a number of other trunk nodes. Hub-spoke suffers from the hub being a potential single point of failure. The hub can be made more survivable, however, if it is located in some form of safe haven away from the tactical area of operations, even if it is only used as a backup to a deployed hub. The meshed topology is best from a jamming resistance point of view because most links are point-to-point, which can be relatively easily deployed so that the directional antennas do not point towards prospective jammer locations. For satellite bearers, hub-spoke is the better topology for resistance to jamming, because the hub can operate in a very tight beam while the spokes operate in the best spot beam available, avoiding the jammer if possible. A jammer in the spot beam provides a significant threat, however, particularly to an FDMA transponder that is fairly easily captured by the jammer.

Economy. The meshed network is the most expensive in trunk nodes and bearers, although redundancy and survivability result. The requirement to deploy nodes so that there is connectivity between the desired nodes places a significant burden on network planners, unlike the hub-spoke network that only requires each spoke node to be able to communicate to the hub. Hub-spoke can be significantly more economical, especially for satellite bearers, providing that survivability issues can be addressed.

6.5.2 Preferred Option

Although the meshed topology is the most expensive in trunk nodes and bearers, it provides the greatest redundancy, capacity, survivability, and flexibility using organic bearers (radio relay and troposcatter). When satellite bearers are required for range extension, however, the hub-spoke topology is more appropriate to restrict internodal connections to a maximum of two satellite links. The preferred option, therefore, is to use a meshed topology for organic radio relay and troposcatter bearers and to overlay a hub-spoke topology for satellite bearers.

6.6 Switching Node Composition

Carrying communications services with differing QoS requirements has always been one of the greatest challenges for telecommunications systems, and is arguably of growing importance. The traditional solution has been to provide a QoS appropriate to one class of traffic (e.g., voice), and force other types of traffic (e.g., non-real-time data) to accommodate what is, for them, a less-than-optimal QoS. Examples of this include the use of packet switches embedded in a circuit-switched architecture in fourth-generation trunk communications systems, and the embedding of data into CNR transmissions in the same manner as voice. The tension created by these differing QoS requirements is at the heart of many key issues for modern networks, such as debates over the relative merits of ATM versus IP technology and the use of voice over IP (VoIP).

Since World War II, voice users have generated the major part of the traffic carried by the tactical communications system. Until recently, the requirement for data was limited to a small number of low-capacity links used to carry text-based message traffic between major headquarters and logistics installations. Therefore, the trunk and CNR subsystems have evolved to support voice traffic, but are not necessarily well suited to carrying the extensive data traffic associated with digitization.

Voice can be carried efficiently by a circuit-switched network and requires low transmission delay (usually less than 150 ms) and low transmission delay jitter, but is relatively tolerant of loss and errors introduced during transmission. Most computer data (including text-based messaging), in contrast, is very inefficiently carried by a circuit-switched network and is intolerant of loss and errors introduced in transmission, but is tolerant of large delay (up to several seconds) and delay jitter. The key to mixing voice and data in the tactical communications system is to be able to provide simultaneously two different qualities of service. For real-time services (e.g., voice and video-conferencing), low delay is required; while for non-real-time services (e.g., data for message traffic), low loss is required.

The aim is to provide appropriate qualities of services for both real-time and non-real-time services across links with capacities ranging from 1 Kbps (CNR) up to 100 Mbps (optic fiber). The tactical communications system should provide a single logical network (i.e., be seamlessly integrated), supporting both connection-oriented and connectionless services. The architecture should aim to use a single network-layer protocol within the network. User data terminals will mostly be configured to use IP as the network-layer protocol. If a different protocol is chosen for the backbone network, translation at the interface is required.

For the purposes of this discussion, we treat the tactical communications system as a network consisting of nodes (capable of performing switching) and bearers (predominantly wireless in the tactical environment) that interconnect the nodes. The keys to the architecture of the network are the means of switching at the nodes and the multiple access techniques on the bearers.

Many options exist—and indeed are implemented in different countries' networks—for efficiently supporting both voice and data, including:

- Providing a circuit-switched voice network with an embedded packet switch to carry data;
- Providing a native ATM network in which all traffic is switched in ATM cells;

- Providing a native IP network in which all traffic (including voice) is routed in IP datagrams;

- Providing an ATM network that carries voice traffic and provides a virtual data-link layer between routers allowing IP data to be embedded in ATM cells (usually known as IP over ATM).

6.6.1   Circuit Switch with Embedded Packet Switch

Tactical trunk systems of the 1980s and early 1990s, such as the U.K. Ptarmigan and U.S. MSE systems, typically provide a digital circuit-switched network, with each circuit having a capacity of 16 Kbps, with trunks having capacities up to 2 Mbps. Each voice channel occupies a single circuit. Data is carried using a secondary packet switch, as illustrated in Figure 6.5, with the circuit-switched network providing the physical layer for the data network. Higher data capacities can be achieved in some systems by the use of channel aggregation, in which a number of circuits connecting two packet switches are combined to provide a single logical channel whose data rate is a multiple of the basic circuit capacity, allowing some additional capacity for engineering and signaling.

The network-layer protocol used in tactical data networks of the 1980s, including Ptarmigan, is X.25, which is suitable for carrying low-rate data that is not delay-sensitive, such as text-based message traffic. It is not suitable, however, for providing the high-rate services required by digitized command posts and logistics installations. Switching in X.25 networks is based on virtual circuits. This provides good support for connection-oriented services, but poor support for connectionless services. Furthermore, the protocols used in modern information systems are based on those used in the Internet, which means that the use of IP as the network-layer protocol for data is a more practical solution for modern information systems that are usually configured to use the protocols of the Internet, primarily IP. High-speed, high-capacity IP routers are commonly available, and could be readily deployed to replace X.25 switches.

The primary disadvantage of aggregating packet data in a circuit-switched network comes from the lack of flexibility in reallocating capacity between voice and data traffic. It is not usually possible to reassign circuits dynamically between voice and data traffic, which can be unacceptably inefficient, particularly where bandwidth is very scarce.

6.6.2   Native ATM

ATM was developed in the late 1980s as a very fast packet-switching protocol. By restricting packets to a fixed length, and establishing virtual circuits

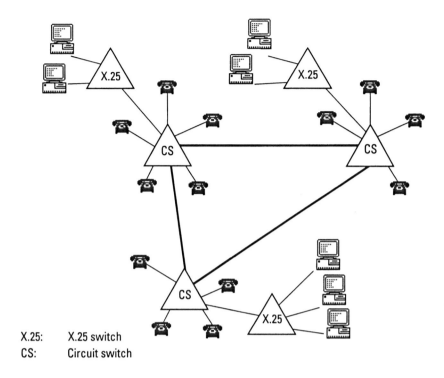

X.25:   X.25 switch
CS: Circuit switch

Figure 6.5 Circuit-switched network with embedded packet switch.

for all data to be carried, very efficient, high-speed switching is possible. A major advantage of ATM is that it is capable of carrying both real-time and non-real-time services simultaneously, while providing an appropriate QoS to both. Furthermore, ATM switches have been available with much higher throughput than for IP routers. ATM's major drawbacks are that a connection must be established before data can be transferred, complicating the use of connectionless services common on the Internet, and that modern information systems are not routinely equipped with suitable interfaces for an ATM network. The architecture of this solution is illustrated in Figure 6.6.

6.6.3 Native IP

One means for overcoming some of the drawbacks of ATM is to use IP to provide both voice and data services, running directly over the data-link layers associated with the network's bearers. This approach has some difficulty providing QoS for real-time services, especially in large networks. Systems providing VoIP are an example of this approach.

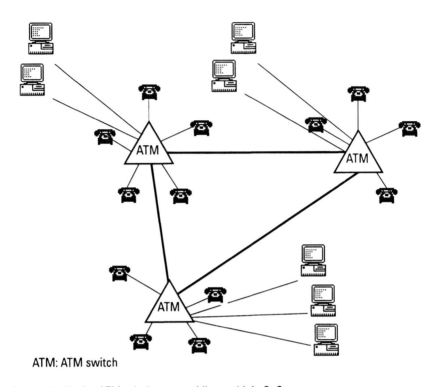

ATM: ATM switch

Figure 6.6 Native ATM solution to providing multiple QoS.

A simple implementation of this concept is shown in Figure 6.7. Each terminal, regardless of the QoS class it supports, is connected to a local router. All datagrams, whether they contain real-time services (voice) or non-real-time services (data) are routed through the network. A number of methods can be used to overcome the fact that IP itself does not support QoS, including connection-oriented prioritization of datagrams, although the difficulty of providing QoS is the major disadvantage of this approach.

An alternative method for using IP in the provision of multiple QoS is shown in Figure 6.8. All data, whether real time or non-real time, is carried in IP datagrams, allowing efficient network support for all traffic. Non-real-time services are routed through the network in largely the same way as for the simple implementation of Figure 6.7. QoS for real-time services is provided by multiplexing these real-time services onto trunks between multiplexers, and providing switching for them in each node.

This dual switching/routing architecture overcomes the difficulty of providing QoS in an IP network, but at the expense of increased complexity in

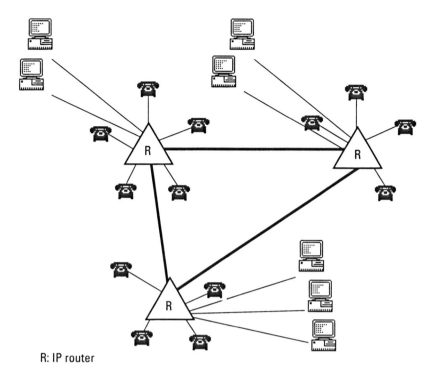

R: IP router

Figure 6.7 Native IP solution to providing multiple QoS.

equipment. This increase in complexity may be minimized by the integration of routers and multiplexers into a single device.

6.6.4 IP over ATM

Combining IP and ATM is seen by some as a means of obtaining the advantages of both, and is usually achieved by carrying IP datagrams inside ATM cells. ATM virtual circuits are established to connect IP routers, effectively providing a virtual data-link layer between the routers, as illustrated in Figure 6.9. This arrangement allows a suitable QoS to be provided for all users, with ATM switches and IP routers often integrated into a single device.

Embedding non-real-time IP traffic in ATM cells that are switched between routers overcomes the inflexibility of embedded packet switch, allowing a flexible reallocation of resources between voice traffic (carried directly in ATM cells) and data traffic (IP datagrams embedded in ATM cells). This approach also supports both real-time QoS (directly in the ATM) with non-real-time QoS (for IP-routed data), in formats that are suitable for

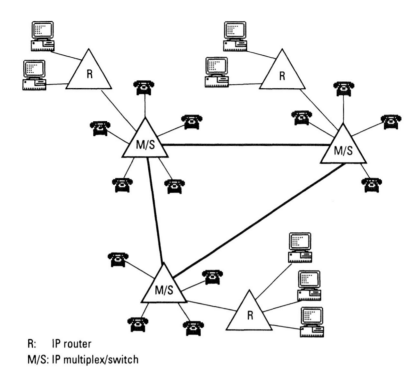

R: IP router
M/S: IP multiplex/switch

Figure 6.8 Alternative use of IP, in which voice is switched rather than routed.

user terminals. Some inefficiency is incurred in the embedding of data into two layers of packetization, each imposing its own overhead for packet headers.

6.6.5   Preferred Option

The preferred approach is to make use of the solutions provided in Figures 6.8 and 6.9. The choice will depend largely on the nature of legacy systems and some preference for particular technology.

6.7   Mobile Access for Trunk Network Subscribers

Mobile remote access is required for trunk network subscribers so that commanders are not confined to their command posts and can command and control on the move. The functionality of the remote access should include full-duplex voice telephony as well as data connectivity at substantial data

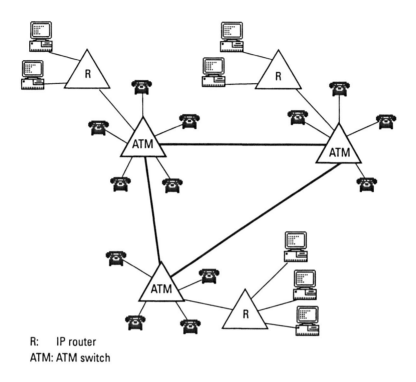

R: IP router
ATM: ATM switch

Figure 6.9 IP over ATM solution for providing multiple QoS.

rates. The remote access channel should be secured to the same level as the remainder of the trunk links in the network. In most major trunk networks this access is currently provided via SCRA access to RAPs that are connected to trunk nodes as described in Chapter 4.

6.7.1 Military Utility

A specialist subsystem to support mobile trunk users might comprise a full-duplex system using any of the three major multiple access mechanisms discussed in Chapter 2 and could utilize any of the current commercial (PCS) or military (SCRA) implementations. It should be noted, however, that current systems are primarily circuit-switched networks that are voice-focused, although commercial systems are rapidly providing data services using packet switching. Wireless LAN technology has too short a range to be seriously considered at this time, although it may be ideal for mobile access for data users within a confined area such as a headquarters.

One of the major limitations with mobile access is the ability to provide ubiquitous coverage of the area of operations. Commercial networks have a considerable investment in fixed infrastructure to provide coverage, and even then there are normally considerable gaps in which it is not cost-effective to provide coverage due to the low density of subscribers. Military networks will always struggle to provide the same extent of coverage, primarily because the forward edge of troops may often only be able to be supported by a base station located within the adversary's area. Additionally, network planners must continually move the base station infrastructure if continuous coverage is to be provided to mobile subscribers moving in accordance with the tactical plan. These issues are discussed in more detail in Chapter 7.

Another major limitation with terrestrial infrastructure for mobile subscribers is the effect of terrain in limiting coverage, and it is likely that comprehensive coverage will only be possible over a force deployed in high-density formations (and even then a significant number of base stations is required to bring mobile coverage down to a company level). More widely dispersed operations will require the range extension of an airborne base station, or use of overlaid (satellite-based) PCS systems.

Capacity.   The mobile access system must have sufficient capacity to support digitized voice and reasonable data rates. Since the subscriber's interface is via a radio channel, the available battlefield spectrum will constrain capacity. Longer ranges (and fewer base stations) will be possible for lower frequencies, so it makes sense for mobile access to operate in the VHF band. Since it also make sense to allocate bandwidth for mobile access on the same basis as other battlefield RF users, 25-kHz channels are appropriate, which will support 16 Kbps encrypted voice or data. It should be noted that the decision to implement mobile access to lower tactical levels will further exacerbate the difficult issue of frequency management in the congested VHF spectrum.

Reliability.   A significant number of base stations is required to provide the overlapping coverage necessary for reliability of coverage, particularly as the supported force disperses. While global coverage of the entire area of operation is not required, planners must have intimate knowledge of tactical plans to ensure that there is sufficient coverage of those areas likely to be required by mobile subscribers. To ensure continuity of coverage, it is also likely that base stations will require the ability provide handoff from one base station to another as the user moves between them, although the user may be satisfied with the much simpler system that requires the user to manually affiliate to the best base station.

Integration. If the tactical mobile subscriber subsystem is a seamless extension of trunk subscriber services, then integration of these users into the wider networks will be provided by the interfaces of the parent tactical trunk subsystem.

QoS. The tactical mobile subscriber subsystem should provide a service of similar quality for real-time and non-real-time services to that which is available to a fixed trunk subscriber.

Communications support for the spectrum of operations. The tactical mobile subscriber subsystem has traditionally provided coverage for mobile users down to battalion level (to company level in some networks) within a relatively dense high-intensity environment. These subsystems become rapidly stressed as the supported force disperses, which will require the range extension of an airborne base station, or use of overlaid (satellite-based) PCS systems. Much greater flexibility is obtained if the mobile access system provides direct access for mobile subscribers. That is, mobile subscribers do not need the base station and can communicate directly when in range of each other. This is a very useful feature for vehicles traveling in convoy during deployment, or when a subunit needs to deploy out of range of the main network. It is also useful because, if two stations are in line-of-sight of each other, they do not have to take up base station bandwidth to communicate. Figure 6.10 illustrates the concept of direct access.

Command and control on the move. A mobile access capability is essential to facilitate command and control on the move. For high-density operations, the base station infrastructure will allow the subscriber to move freely within the coverage of the trunk network. As the force disperses, however, network planning becomes much more difficult and the users may have to accept that there will be instances when coverage is not available. Again, a direct-access facility will greatly ease this issue as mobiles can communicate with any other mobile that is within range, regardless of whether the base station is available or in range. While access to the trunk network is not available during direct access, at least communications would be possible between nearby units, or from a commander to subunits.

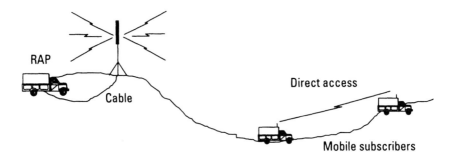

Figure 6.10 Direct access for mobile subscribers.

Organic, minimum-essential communications.   Base stations should be organic to the tactical trunk subsystem. For movement and administration, base stations may also be considered to be organic to various elements of the deployed force.

Security, jamming resistance, and economy.   Adoption of commercial PCS standards may provide a more economical solution, although the use of PCS terminals may be a cause for concern from the point of view of security and jamming resistance—see Chapters 4 and 7 for more detailed discussion.

6.7.2 Preferred Option

Mobile trunk network access is essential for headquarters down to company level. Additionally, direct access is a critical requirement of any mobile access solution. Current SCRA systems support these requirements to varying degrees and all systems will be improved by the incorporation of the technologies developed within commercial PCS systems. The most significant changes will be in the larger number of base stations required to support the lower level of access and in the addition of range extension provided by the tactical airborne subsystem or satellite-based PCS from overlaid systems.

6.8 Interfaces to the Tactical Trunk Subsystem

The tactical trunk subsystem is required to interface to other components of the tactical communications system as well as to the strategic communications system and to the overlaid communications system. These generic interfaces are illustrated in Figure 6.11. Detailed descriptions of the interfaces are provided in Chapter 10.

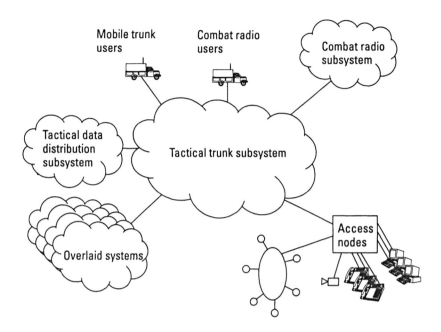

Figure 6.11 Tactical trunk subsystem interfaces.

6.9 Summary

The architecture of the tactical trunk networks of most modern armies is well placed to accommodate the requirements of the digitized battlefield. The architectural components of switching nodes, access nodes, bearers, and mobile subscriber facilities are still required in future systems. The internal structure of these components must be modified, however, with the most change needed to the switching fabric within the subsystem. Additionally, the components must be modified to accommodate the greater mobility required by users, as well as the dispersion necessary to support the spectrum of operations.

Perhaps the largest architectural change is in the interface requirements for the subsystem. Currently, the trunk subsystem is only required to interface to other trunk subsystems as well as provide limited interface to CNR users (CNRI). Future trunk subsystems will take the prime role in integrating all of the other subsystems in the tactical communications system—these issues are discussed in much more detail in Chapter 10. Chapters 7 through 10 continue to take a more detailed look at the combat radio, tactical data distribution, and tactical airborne subsystems.

Endnotes

[1] Rice, M., and A. Sammes, *Command and Control: Support Systems in the Gulf War*, London: Brassey's, 1994, p. 25.

[2] Frater, M., and M. Ryan, *Electronic Warfare for the Digitized Battlefield*, Norwood, MA: Artech House, 2001, Chapter 3.

7

The Combat Radio Subsystem

7.1 Introduction

Military requirements for communications with fully mobile infrastructure have been met traditionally by CNR, which is the primary means of exercising command and control at brigade level and below. CNR combines the advantages of simplicity and flexibility with the ability to provide the all-informed communications that are essential for the close coordination of all-arms tactics in mobile operations.

The use of radio on the battlefield began in World War I as an alternative to line as a part of trunk communications, which avoided the laying of hundreds of miles of cable to support major offensives. Examples included the connection of observation posts to artillery batteries. Radio sets—and particularly their antenna systems—were initially too large to be of any great use to the infantry. However, as sets and antennas reduced in size, they began to be employed to form artillery-infantry nets and infantry-armor nets and became more useful in mobile operations.

By the end of World War II, CNR had become an important means of communications for infantry and other arms. Technical developments since that time have been evolutionary rather than revolutionary, and the tactical use of CNR has remained largely unchanged. The major difference is in the ability to pass data, although most CNR systems are still analog radios and are not well placed to cope with the expansion in the volume of data expected in the next few years.

The main advantages of CNR are that it is simple, flexible, robust, and easily deployed, providing the fully mobile infrastructure that is required for the command and control of combat troops. The position of the CNR subsystem in the range/capacity/mobility trade-off, sacrificing capacity to maximize range and mobility, is shown in Figure 7.1.

The requirement to provide a battlefield communications system with a fully mobile infrastructure will continue for the foreseeable future, especially at battalion and below, where it will continue to provide the primary means of command and control, and to a lesser extent at higher levels where it is required to supplement the array of trunk communications systems available to commanders and staff. This fully mobile system, which we call the combat radio subsystem, is the subject of this chapter. We begin by

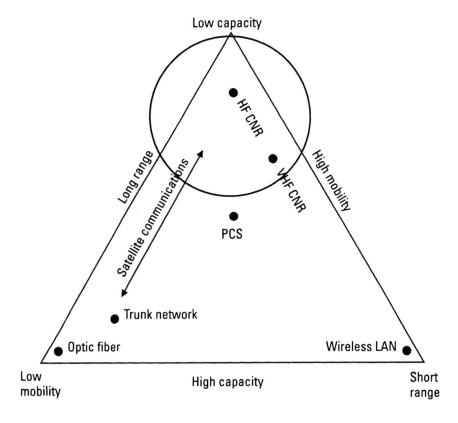

Figure 7.1 The place of the combat radio subsystem in the range/capacity/mobility trade-off.

examining the basic requirements for the combat radio subsystem, and use this analysis to compare a range of currently available mobile communications technologies.

7.2 Key Architectural Drivers

The basic requirements of the combat radio subsystem are based on the requirements for the tactical communications system detailed in Chapter 5, but with refinements to meet the need for a battlefield communications system with fully mobile infrastructure.

Range/capacity/mobility trade-off. Each user community supported by the combat radio subsystem must be provided with a minimum of one shared voice channel. Simultaneous transmission of voice and data may not be possible. Having met this minimum capacity, increased mobility and range have higher priority than increased levels of capacity, as shown in Figure 7.1.

Command and control on the move. The primary purpose of the combat radio subsystem is to support command and control of combat troops. User terminals and network infrastructure must be capable of operation while on the move without stopping. This requires that either there is no ground-based infrastructure or that this infrastructure is fully mobile. The requirement for command and control on the move makes line unsuitable as a sole means of communications; it favors, however, the use of radio with omnidirectional antennas. Radios and terminals must also be small and robust with low power requirements if they are to provide integral support to motorized and mechanized forces and especially light-scale forces. Requiring users to type all information into a computer is not acceptable; voice communications must be provided as a primary means, with data supported when required.

Multiple access. The spectrum available for military use is not likely to expand, while the number of systems that make use of the electromagnetic spectrum increases constantly. Sharing of the electromagnetic spectrum among users is required. Possible multiple access techniques include FDMA, synchronous TDMA, CSMA, and CDMA or some combination of these basic techniques. In current systems, multiple access is achieved by grouping stations into nets. Each net operates in a single-frequency, half-duplex mode, with different nets being assigned different frequencies.

Support for the chain of command.   Support for the chain of command requires that communication between a commander and subordinates is achieved with maximum efficiency, and that each level of the command hierarchy (at least from brigade down to platoon) is provided with integral combat radio communications assets and the capability to manage these assets. This is achieved in current systems by hierarchical nets that follow the chain of command. Support for the chain of command also requires that ground-based equipment used for communications within a unit or subunit is integral to that unit or subunit. The combat radio subsystem should not constrain the locations of headquarters or other elements. This applies both while on the move and in static locations. A commander should also be able to alter command arrangements within his or her formation or unit without having to fundamentally restructure the combat radio subsystem.

QoS.   The combat radio subsystem is required to carry a wide variety of traffic. Traditionally, voice is the major form of traffic; however, the demand for various types of data is increasing. In modern networks, it is more useful to characterize them by the requirements that each type of data has for services across the network. There are two main types of traffic: those that require real-time services (predominantly video and voice) and those that require non-real-time services (computer-to-computer transfer). Both voice and data communications are required, therefore, although not necessarily at the same time or using the same equipment. Ideally, real-time and non-real-time (i.e., voice and data) communications should be seamlessly integrated, using a single piece of equipment. In practice, a data network that falls back to a group of hierarchical chain-of-command voice nets may be acceptable. A major disadvantage of separate equipment for voice and data may be additional weight.

Multicast capability.   Traditional CNR provides an all-informed voice capability that is ideally suited to the coordination of all-arms tactics. This requirement for all-informed voice capability is not likely to decrease. With the introduction of a data capability into the combat radio subsystem, a corresponding multicast capability for data is also required.

Flexibility.   The combat radio subsystem should support infantry, motorized infantry, and mechanized infantry operating over the spectrum of operations ranging from high-level conflict to dispersed peacekeeping and low-level counterinsurgency operations. Requirements for training of operators, network managers, commanders, and staff will dictate that there should be minimal differences between the equipment used for different types of operations. Any controls must be able to be used while the operator is wearing gloves, either for protection from the cold or in an NBC environment.

Seamless connectivity.   Seamless connectivity should be provided both within the combat radio subsystem and between this subsystem and the tactical trunk subsystem. This may require the rebroadcast of data within the combat radio subsystem or carriage by the tactical trunk subsystem of some data whose source and destination both lie in the combat radio system. In practice, seamless connectivity is a higher priority for data than for voice and, in fact, may be feasible only for data. Typically, the interface to the tactical trunk subsystem would be part of that subsystem.

Security.   In the past, the use of secure communications has been limited by the weight and bulk of encryption devices, and procedural restrictions on the circumstances under which such devices could be used. With the increased use of secure communications equipment in the commercial world and the availability of low-cost, integrated communications security for combat radio, secure communications should now be provided by the combat radio subsystem at all levels.

EP.   EP is required to provide LPD and resistance to jamming. This might include passive EP, such as terrain shielding or the use of directional antennas, and active EP, including DSSS and frequency hopping. Increasingly high levels of integration of EP are becoming possible due to advances in integrated-circuit technology.

Minimal mutual interference.   Equipment forming the combat radio subsystem should be capable of operation in close proximity with other equipment of the same or similar type. The major practical implication is that frequency management is required to ensure that closely spaced frequencies and certain harmonic combinations are not used in the same location.

Power source.   Many types of operations, such as dismounted infantry, will require operation of the combat radio subsystem on battery power for extended periods. A requirement for regular replenishment of batteries decreases the flexibility of a force, and also increases the burden on logistics. U.S. doctrine, for example, suggests that a combat radio should be capable of continuous operation for 24 hours without replenishment of batteries. While this is not achieved by current systems, recent advances in battery technology and power management techniques, both largely due to the commercial cellular telephone, have the potential to provide such extended operations.

Operation in all geographic and climatic conditions.   Combat radio equipment is required to operate in a range of conditions beyond those normally expected for commercial equipment. This includes immersion in water, a temperature range of −20°C to +70°C, and withstanding extremes of vibration, shock, pressure, and humidity [1].

7.3 Multiple Access

In the combat radio system, many stations transmit RF energy onto a single channel. A multiple access technique is required to share the electromagnetic spectrum between these stations.

The requirement for seamless connectivity means that the combat radio subsystem should operate as a single logical network. This requirement does not, however, dictate that there should be a single physical network. An examination of traffic flows in a typical network will show that users tend to form groups within which large amounts of data are exchanged, but with much smaller amounts of external traffic. It is for this reason that the concept of a subnet is commonly used in computer networks. The nets typically used in the combat radio subsystem also have this property.

For the purposes of this discussion, we define *a user community* as a group of stations among which the combat radio subsystem allows direct communications, assuming that the radio operating range permits, without passing through a bridge or router. We do not impose any particular structure of these user communities; indeed, they need not follow current doctrine of CNR nets. In the extreme of aggregation each station in the network may be part of a single, all-encompassing, user community. This definition can also encompass a circuit-switched cellular telephone system, for which a user community is formed by a base station and the set of mobile units that can communicate directly with it.

The implementation of a multiple-access scheme requires the resolution of two key issues: the mechanism for controlling access by users to the bandwidth assigned to the user community and the mechanism for limiting the interference between different user communities. The utility of the various multiple-access techniques is discussed next.

Synchronous TDMA.   In synchronous TDMA, a fixed-length, periodic time slot is allocated to each transmitter. Timing synchronization requires a common time reference (e.g., GPS time), a central control station, or regular transmissions from all stations. Allocation of time slots can be by fixed assignment or on-demand, the latter method providing much greater flexibility but requiring a central control station. The advantages of TDMA are that it is relatively easy for one station to monitor transmissions in all time slots, it is possible to use close to 100% of the available channel capacity if there is either a single transmitter or single receiver, there is no loss of data due to one station overtransmitting another, and there is a fixed upper bound on delay. The disadvantages are that timing synchronization between stations is required, guard intervals reduce channel capacity when multiple transmitters and multiple receivers are used, and allocation of capacity between transmitters is relatively inflexible.

CSMA.   CSMA techniques are a form of asynchronous TDMA, in which there are no fixed time slots. A station wishing to transmit first checks that no other station is currently transmitting. The advantages of CSMA are that it is relatively easy for one station to monitor all transmissions on the channel, no central control station is required, and the allocation of channel capacity is very flexible. The disadvantages of CSMA are that the best throughput that can be achieved in a typical operation is approximately 50% of the available channel capacity, data is lost due to one station overtransmitting another, and there is no fixed upper bound on delay. The efficiency of CSMA can be improved by the introduction of a control station that assigns channel capacity to stations on demand.

FDMA.   FDMA allocates a portion of the electromagnetic spectrum to each transmitter, which can transmit in its allocated channel all the time. Allocation of frequencies can be fixed or on-demand. The advantages of FDMA are that no central control station is required unless capacity is demand-allocated, close to 100% of the available channel capacity can be used, there is no loss of data due to one station overtransmitting another, and there is no delay introduced by the channel. The disadvantages of FDMA are that the allocation of capacity between transmitters is relatively inflexible and it is relatively difficult for one station to receive data from more than one transmitter. While FDMA is employed to separate user communities, it is rarely used as a multiple-access technique within a user community.

Frequency hopping.   As a multiple-access technique, frequency hopping is a form of FDMA. Each transmitter can transmit 100% of the time. As the number of transmitters is increased, the proportion of hops on which over-transmission occurs also increases. The advantages of frequency hopping as a multiple-access technique are that no central control station is required, the allocation of capacity is very flexible, and the total transmitted data rate can be in excess of 100% of the available channel capacity, as long as a high error rate can be tolerated. The disadvantage of frequency hopping as a multiple-access technique is that some data is lost due to overtransmission.

CDMA.   CDMA—also known as DSSS—allows a wideband channel to be shared by a number of narrowband sources by spreading their transmissions over the whole band. By using a different spreading sequence for each transmitter, multiple access is achieved. The advantages [2] of CDMA are that no central control station is required unless capacity is demand-allocated, close to 100% of the available channel capacity can be used, there is no loss of data due to one station overtransmitting another, and there is no delay introduced by the channel. The disadvantages of CDMA are that the near-far effect makes it infeasible to have more than one transmitter and more than one receiver operating simultaneously, the allocation of capacity between transmitters is relatively inflexible, and it is relatively difficult for one station to receive data from more than one transmitter.

Time hopping.   Data may be encoded in variations in the length of intervals between the transmission of very short impulses. These impulses may be as short as 1 ns, giving a total transmission bandwidth of 1 GHz. This approach is used in ultra-wideband radio (UWB), also known as impulse radio. The advantages of UWB radio are that it has the potential to completely remove the need for frequency management as it is now practiced, it allows for very flexible allocation of channel capacity to a transmitter, and it makes possible extremely simple digital receivers. The major disadvantage is the high level of interference caused to all conventional communications systems whose bandwidth is shared by the UWB system.

As well as providing multiple access, frequency hopping, time hopping, and CDMA may also be used to provide LPD.

For conventional CNR, each user community forms one net. Each net operates on a single frequency, with FDMA being used to share the electromagnetic spectrum between nets. Within a net, a form of CSMA is used to share the channel capacity between stations on the net. For voice networks, CSMA takes the form of voice procedure [3].

7.3.1 Multiple Access Within a User Community

Synchronous TDMA. As long as a common time reference is available (possibly transmitted by a central control station) and the flexible allocation of channel capacity is not required, synchronous TDMA can be a very effective means of providing multiple access within a net. TDMA is used by GSM cellular telephone networks, EPLRS [4] and TADIL-J (Link-16). The major disadvantage of TDMA for the combat radio subsystem is the power consumption resulting from transmissions required to maintain synchronization. With current battery technology, it seems likely that at least the central control station must be vehicle-mounted.

CSMA. CSMA is the access technique used by current CNR, for both analog and digital voice. By shortening transmissions through digitization, throughput in a CSMA system may be significantly increased. The major remaining drawback of CSMA is that it is unlikely that a throughput higher than approximately 50% of channel capacity can be achieved for data.

FDMA. As long as the flexible allocation of channel capacity is not required, FDMA can be used to provide multiple access within a net. The major drawback is the difficulty of monitoring more than one channel at a time, which works against the requirement for all-informed communications.

CDMA. Because of the near-far effect, CDMA is not suitable for use as a multiple-access technique except where there is only a single transmitter or a single receiver. This is achieved in mobile telephone networks because all transmissions either emanate from or are destined for a base station. In many military applications, it is usually the case that all stations are required to be able to transmit and to receive transmissions from all other stations.

Frequency hopping. The use of frequency hopping to provide multiple access within a net leads to an inflexible allocation of channel capacity and usually makes monitoring of more than one transmitter infeasible. For this reason, frequency hopping is not usually used for multiple access within nets.

Time hopping. Time hopping provides a very flexible allocation of capacity to transmitters. It is likely, however, that the complexity of a receiver that attempts to monitor n transmissions is likely to be n times that of a single channel receiver, making time hopping unsuitable for providing multiple access within a user community.

7.3.2   Multiple Access Between User Communities

Synchronous TDMA.   The requirement for a synchronization of timing usu-ally makes the use of TDMA for providing multiple access between user communities infeasible. This is because it would be unusual for all stations to be able to monitor the transmissions of all other stations and because large guard intervals would be required to compensate for the difference in trans-mission path lengths. Where the total number of users in a group of user communities is not too large, TDMA may be used as an adjunct to another multiple-access technique (such as FDMA).

CSMA.   Efficient operation of CSMA requires a small number of transmit-ters and that all transceivers can hear all other transmitters. This is rarely the case across a wide area, making CSMA infeasible as a multiple-access tech-nique between user communities.

FDMA.   FDMA is commonly used for providing multiple access between communities. In fact, it is a useful way of separating user communities.

CDMA.   The near-far effect usually makes the use of CDMA for providing multiple access between user communities infeasible.

Frequency hopping.   Frequency hopping can provide an effective means of multiple access between user communities, as long as deletions due to over-transmissions can be tolerated. This is usually acceptable for secure voice, where up to one-third of the data can be lost before speech becomes unintel-ligible. These deletions may be acceptable for data if suitable error protection is provided. In this case, however, error-correction data leads to a loss of overall throughput.

Time hopping.   The provision of multiple access between user communities can make use of all the advantages of time hopping, providing for very flexi-ble allocation of capacity, the removal of the need for frequency manage-ment, and the possibility of using simple, digital receivers.

7.3.3   Summary

Of all the technological options for the provision of multiple access, only a relatively small number are suitable as part of the combat radio subsystem. Within a user community, synchronous TDMA, CSMA, CDMA, or time hopping may be used. Between user communities, FDMA and time hopping are likely to be the only feasible options.

7.4 Candidate Solutions

In this section, candidate solutions for the provision of a fully mobile tactical communications system are examined. The following technologies are candidate solutions: CNR with a capability to pass digital data, packet radio, a repeated TDMA system, cellular telephone/PCS, and trunked radio.

7.4.1 Data-Capable CNR

Traditional CNR provides a single voice channel. Multiple access is provided within nets by CSMA, and between nets by FDMA. CNR is used in the HF (2–30 MHz), VHF (30–88 MHz), and UHF bands. The modulation scheme for HF CNR is usually SSB, with a channel bandwidth of 3 kHz; for VHF and UHF, FM is commonly used with a channel bandwidth of 25 kHz or 50 kHz.

Data modems have been available for CNR for more than 20 years. A single HF CNR channel can carry up to approximately 2.4 Kbps, while a VHF or UHF CNR channel can carry 16 Kbps with a channel bandwidth of 25 kHz. CSMA is used to control multiple access. In areas where automatic control is provided, an operator can prepare a message and have this message transmitted asynchronously. Many in-service analog VHF radios provide a 16-Kbps data channel, principally to meet the requirement to transmit encrypted voice.

Most in-service CNRs have been designed to participate on a single-frequency, half-duplex voice net. To provide a data net using such radios requires external equipment in addition to the data-capable CNR to provide for the switching or routing of data. Most current-generation combat radio equipment uses this approach.

7.4.1.1 Military Utility

Range/capacity/mobility trade-off. High mobility is possible, especially where radios are fully integrated with cryptographic equipment and data modems. The major limitation is the throughput that can be achieved for data, which is typically no more than 4 Kbps for VHF CNR and 300 bps for HF CNR. The throughput is limited by both the properties of CSMA and the requirement to use a high level of forward error correction to obtain an acceptable error rate.

Command and control on the move.   Command and control on the move are supported by VHF and UHF CNR, and HF CNR using ground-wave propagation. Voice is the primary mode of operation. HF sky-wave propagation is not usually possible without the erection of large, fixed antennas. Efficient operation for data requires that direct communications be possible between all stations on the net, that is, the net must not be fragmented. This depends strongly on terrain. Retransmission stations can be used to minimize the impact of fragmentation, although this can be difficult on the move.

Multiple access.   Multiple access is achieved by a combination of CSMA within a net and FDMA between nets, providing a high level of flexibility.

Support for the chain of command.   Communications follow the chain of command. The location of headquarters is constrained to some extent by the requirement for direct communications between all stations on the net. This is accentuated by the use of data communications.

QoS.   Voice and data can be supported with a single piece of equipment. This is common for VHF CNR, although a separate modem is often required for carrying data on HF CNR. A net can operate in either a data mode or voice mode. A data net may drop back to voice operation when an operator presses the pressel switch, providing an appropriate QoS to each service.

Multicast capability.   All-informed voice and a multicast capability for data are to be supported.

Flexibility.   Similar equipment can be used to support troops across the spectrum of operations. Ancillary equipment, such as RF amplifiers and antenna-tuning units, is used to provide increased functionality for vehicle-mounted systems. Weight can be a limitation for foot-mounted operations, especially where separate equipment (and therefore separate batteries) is required for secure operation and EP.

Seamless connectivity.   External equipment is required to provide connectivity between nets, or to other subsystems of the tactical communications system. Voice connectivity to the tactical trunk subsystem may be provided by a CNRI.

Security. Secure operation for voice and data has traditionally been provided by external cryptographic equipment such as the KY-57 VINSON and KY-99A MINTERM. However, more recent designs, including recent versions of SINCGARS, have an integral cryptographic capability.

EP. Frequency hopping and DSSS may be supported, either by external equipment or internal options (e.g., in recent versions of SINCGARS).

Minimal mutual interference. As long as proper frequency management is carried out, preventing the use of closely spaced frequencies and certain harmonic combinations in the same location, CNR equipment provides immunity to interference from closely spaced transmitters.

Power source. CNR can be powered for extended periods by batteries. In areas where external ancillary equipment is used, such as for secure communications, this equipment usually requires separate batteries.

Operation in all geographic and climatic conditions. CNR systems are typically designed to meet the full range of battlefield conditions.

7.4.1.2 Current Availability

Since the need for networked data-capable radios is common in most modern armies, a number of systems are available off-the-shelf.

7.4.2 Packet Radio

The major disadvantage of data transfer over CNR is that there is no support for automatic rebroadcast of data in fragmented nets. This can be overcome, at the expense of added complexity, by the use of packet radio [5].

A packet radio system uses the same net structure as conventional CNR, as illustrated in Figure 7.2. Because of the presence of the left-to-right links and the collocation of stations on multiple nets within headquarters, this net structure is highly meshed, which can provide a high level of redundancy as long as procedures exist for taking advantage of this meshing. Packet radio takes advantage of the collocation of stations by providing an *internet bridge* between nets at each of these locations. These internet bridges allow automatic delivery of messages addressed outside a particular net and

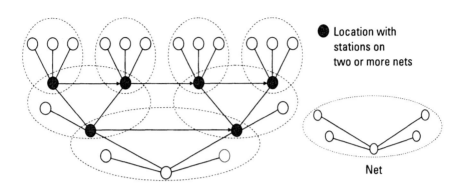

Figure 7.2 Illustration of the operation of the tactical internet, showing locations of possible internet bridges.

alternate routes around parts of the network that are unavailable. The overall structure of the network is called a *tactical internet*.

Within a net, messages are automatically relayed if the net becomes fragmented. This process is known as *intranet rebroadcast*. This is illustrated in Figure 7.3. With a raw capacity of 16 Kbps for each net, it is likely that a throughput of approximately 1 to 3 Kbps will be achieved. In reality, internet rebroadcast may be limited to two or three hops before the throughput of the network becomes unacceptably low.

The internet bridging and intranet rebroadcast functions of packet radio are applicable only to data. Voice traffic is accommodated by allowing the packet radio network to drop back to a chain-of-command, hierarchical network when an operator depresses the press-to-talk switch. The secure-voice mode therefore has priority over the data mode. In the voice mode, operation of nets is identical to that of conventional secure-voice radio nets. The data mode is resumed immediately when transmission of voice ceases. Voice operations in fragmented nets are similar to those of conventional CNR nets.

Secure communications can be provided using on-line, military-grade encryption. By the implementation of partially programmable encryption systems, new algorithms can be supported side-by-side with those from the previous generation of secure communications equipment. Packet radio may be used in conjunction with frequency hopping and free-channel search to reduce vulnerability to interference and jamming in a hostile electromagnetic environment.

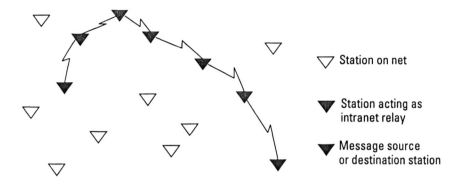

Figure 7.3 Illustration of operation of the tactical intranet, showing relay stations.

The advantages of the tactical internet are:

- Data can be passed automatically between any two locations on the battlefield, without manual retransmission at net boundaries.
- CNR operates without support of external network devices.
- Messages can be routed around failed parts of the network, because the transmission path is not fixed.

The disadvantages of the tactical internet are:

- Because of the low data rates available (1–3 Kbps), the network can easily become congested.
- It may be necessary to limit the number of intranets across which a message is passed to prevent the whole network becoming congested in the event of failure of one part of it.

The advantages of the tactical intranet are:

- Efficient operation of a fragmented net is possible.
- Rebroadcast is provided with a single radio at each site.
- Adversary intercept of transmissions is hindered by use of lower power levels than would be required for direct communication between all stations on a net.
- Traffic analysis of a net may be made more difficult, with many transmissions being rebroadcasts rather than new messages.

The disadvantages of the tactical intranet are:

• An operator has less control over the transmission a radio than in traditional hierarchical nets.

• Because of the low data rates available (1–3 Kbps) and the requirement for a multiple access protocol, the network can easily become congested.

• Operation with highly fragmented nets does not appear to be feasible, with intranet rebroadcast limited in practice to one or two hops.

7.4.2.1 Military Utility

Range/capacity/mobility trade-off.   Mobility is similar to conventional CNR. Each net supports a single voice channel. In data mode, a total capacity of 1 to 3 Kbps is possible on each net. This capacity is similar to what can be achieved with conventional CNR and may not be sufficient for some future digitization requirements, such as real-time situational awareness.

Command and control on the move.   Command and control on the move are supported in a similar manner to data-capable CNR. Provision of intranet rebroadcast simplifies the maintenance of connectivity in a net. Data communications on the move may be limited by the ability of soldiers to operate terminals while moving.

Multiple access.   Multiple access is provided in the same way as for traditional CNR; FDMA separates nets, while CSMA is used to share a single channel between stations on a net.

Support for the chain of command.   The net structure for packet radio would typically be identical to that used for traditional CNR. Constraints on the locations of headquarters are reduced by the removal of the requirement for direct communications between all stations in an intranet and by the ability to have a station other than a headquarters act as the bridge to another net.

QoS.   Voice and data are usually provided as separate modes of operation, providing different QoS to each.

Multicast capability.   The multicast capability is essentially the same as for data-enabled CNR, but with the added reach offered by the intranet rebroadcast.

Flexibility.   Packet radio operates effectively across the spectrum of operations.

Seamless connectivity.   Seamless connectivity is provided for data communications within the packet radio system by means of intranet rebroadcast and internet bridging. External interfaces are still required for interface to external communications systems, such as the tactical trunk subsystem. These interfaces would not usually be provided as part of the combat radio subsystem because of their associated weight and management requirements.

Security.   Encryption for both voice and data is usually provided using integral cryptographic devices. This is advantageous in terms of the performance offered at the reduction in weight compared to the use of separate cryptographic equipment, but may impose significant procedural and administrative constraints.

EP.   Intranet rebroadcast permits lower transmission power to be used in some applications, reducing the probability of detection and interception. Because packet radio equipment is of more recent design than CNR, EP (e.g., frequency hopping) is often built-in rather than requiring separate appliqué equipment.

Minimal mutual interference.   As long as proper frequency management is carried out, preventing the use of closely spaced frequencies and certain harmonic combinations in the same location, packet radio provides immunity to interference from closely spaced transmitters.

Power source.   Packet radio can be powered by batteries, although extensive use of the intranet rebroadcast is likely to reduce battery life significantly.

Operation in all geographic and climatic conditions.   Packet radio systems, being specifically designed for military applications, are typically designed to meet the full range of battlefield conditions.

7.4.2.2 Current Availability

The SINCGARS ASIP radio, used with the external internet controller to provide the packet radio capability, is in service with some units of the U.S. Army and has recently also been procured by the New Zealand Army. The British Army is aiming to procure a similar capability as part of its Bowman program.

7.4.3 Ad Hoc Networks

In simple terms, an *ad hoc network* uses techniques similar to the intranet rebroadcast of a packet radio network to provide connectivity throughout the network. The use of ad hoc networks in the combat radio subsystem has the promise of removing the need for communications to follow the chain of command, reducing the requirement for detailed planning of the combat radio subsystem, and potentially greatly easing the task of providing communications to fast-moving forces. Traditionally, the effective planning for the provision of rebroadcast to such forces has been very difficult to achieve. For example, despite new doctrine to make use of available assets for the advance, U.K. forces in the Gulf War found it impossible for rebroadcast vehicles to keep up with main battle tanks and APCs [6].

In an ad hoc network, stations cooperate to build the network. Stations communicate using a common wireless channel. Each station can communicate directly with one or more of the other stations in the network, but it is unlikely that any one station can communicate directly with all of the other stations. Stations on the network are therefore required to act as relays. Data is carried through from source to destination by being passed from one relay to the next. Each station maintains a list of the stations to which it can directly communicate. Connectivity information is built up and distributed by each station [7].

In the example network shown in Figure 7.4, *B* can communicate directly with *A* and *G*. *B* may send data to *E* via the path *BGE* or the path *BACE*. Each of these paths would have an associated cost, which may be as simple as the number of hops involved. *B* would choose the least-cost path, transmitting the data over the first hop. The relay station (say, *G*) then transmits the data over the next hop, with this process continuing until the data reaches its destination.

In larger ad hoc networks, stations may form themselves into clusters. A small number of stations may then take on the role of communicating between clusters, possibly using higher transmission power to do so. The forming of clusters helps to maximize frequency and battery life reuse by

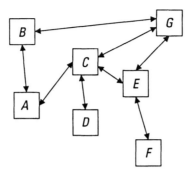

Figure 7.4 Example of connectivity within an ad hoc network.

minimizing transmission power. In the example in Figure 7.5, *E* and *F* have taken on the role of intercluster communication.

An ad hoc network may be integrated with a wider network by one of the stations on the ad hoc network acting as a gateway. In the context of the tactical communications system, this interface point must be defined. This issue is addressed further in Chapter 10.

The major utility of an ad hoc network in the tactical communications system is the fact that the network is formed by the terminals, without the requirement of a specific infrastructure to be deployed.

7.4.3.1 Military Utility

Range/capacity/mobility trade-off. The ad hoc network provides very high mobility by removing the need for each station in a user community to stay within range of all other stations. This is only achieved where a sufficiently high density of emitters is available.

Command and control on the move. The ad hoc network is likely to be oriented toward data. Support for command and control on the move probably requires that the system be able to fall back to a chain-of-command, voice network.

Multiple access. Multiple access is controlled by the ad hoc networking protocols. The efficiency of multiple access is likely to be lower than for a planned system, such as CNR or packet radio. This reduced efficiency may not be acceptable in a congested electromagnetic environment.

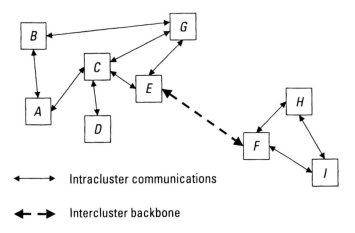

Figure 7.5 Example of clustering in an ad hoc network.

Support for the chain of command.   The structure of the ad hoc network adapts to the physical disposition of a force. So long as the connectivity of the network is maintained, communications supporting the chain of command are possible. Maintenance of the structure of the network in dispersed operations may be difficult without additional network infrastructure. Because the combat radio subsystem must be provided as integral assets at a number of levels of hierarchy, relevant network management assets must also be provided at each level.

QoS.   Provision of a suitable QoS for data is easily achieved in an ad hoc network; provision of QoS for real-time voice is unlikely to be possible except for stations that can communicate directly. This difficulty may be overcome in the same manner as for packet radio, allowing operation to drop back to a hierarchical, chain-of-command, voice network when the operator depresses the press-to-talk switch.

Multicast capability.   An ad hoc network naturally supports multicast, using similar techniques to packet radio. The provision of multicast, however, is only efficient where all destination stations are in direct communication with the transmitting station.

Flexibility.   The ad hoc network relies on having a sufficient density of stations for its operation. This is likely to be problematic in dispersed operations, unless some form of planned, rebroadcast infrastructure is supplied. The use of the airborne subsystem greatly reduces these difficulties.

Seamless connectivity. An ad hoc network provides seamless connectivity as a natural part of its operation, at least for non-real-time services. For real-time services, the ad hoc network has the same difficulties with efficiency of operation as data-enabled CNR and packet radio.

Security. An ad hoc network is likely to have an integral security architecture. However, the quality of encryption provided in commercial ad hoc networks may not be acceptable.

EP. An ad hoc network provides some protection from adversary intercept by removing the direct connection between network structure and chain of command. Commercial ad hoc network standards are likely to be very vulnerable to jamming.

Minimal mutual interference. An ad hoc network should provide its own frequency management, minimizing the need for planning in this area.

Power source. The life of batteries for man-portable systems in an ad hoc network is likely to be a major concern, especially for stations providing intracluster backbones. This problem may be partially overcome by the designation of specific stations (perhaps those mounted in vehicles) to carry out this role. The drawback of this is that it limits the flexibility of the system by effectively requiring data between clusters to pass through a base station. The disadvantages of this solution are discussed further in Section 7.4.5.

Operation in all geographic and climatic conditions. Commercial ad hoc networking equipment may require ruggedization before being suitable for use across the full range of geographic and climatic conditions.

7.4.3.2 Current Availability

While it is the subject of much research and, indeed, funding from organizations such as the Defense Advanced Research Projects Agency (DARPA), the ad hoc network technology is not yet sufficiently mature to appear in fielded systems.

7.4.4   Repeated TDMA

A repeated TDMA network overcomes the inefficiency of packet radio by requiring all stations to transmit in allocated time slots. All network timing and operation are controlled by a common time reference, usually transmitted by a net control station. A number of repeated TDMA radios are currently in service in the United States, including the EPLRS radio and TADIL-J (Link 16). A repeated TDMA system is usually intended to operate with much larger user communities than a typical CNR net, possibly including all of the vehicles in a brigade in a single user community. The capacity of the system, which may be up to 500 Kbps, is shared between these users. Under the direction of the net control station, rebroadcast is provided between stations needing to communicate, but who are out of direct communications range. The maximum range over which a repeated TDMA system can operate is limited by the requirement to provide guard intervals between TDMA slots.

Repeated TDMA systems are discussed in more detail in Chapter 8, dealing with the tactical data distribution subsystem.

7.4.4.1   Military Utility

Range/capacity/mobility trade-off.   While the total capacity of a repeated TDMA system is likely to be much higher than that of a CNR net, this capacity tends to be shared among a larger user community. It is therefore unlikely for a repeated TDMA system to have sufficient capacity to support the voice communications requirement of its user community. The range of individual stations is limited by their mobility, but this is offset by the ability to provide automatic repeating for stations outside the range for direct communications.

Command and control on the move.   Repeated TDMA systems are capable of operating at full capacity while stations on the net are moving. Manpack operations may be limited by current requirements for the network control stations to be vehicle mounted.

Multiple access.   To the user, a repeated TDMA system appears like a circuit-switched network that permits some circuits to be used for all-informed nets. This is achieved using a combination of TDMA, FDMA and CDMA, and provides efficient use of the electromagnetic spectrum.

Support for the chain of command. A repeated TDMA system provides a meshed network in which each radio set is able to provide retransmission for stations that are beyond line-of-sight. This retransmission is usually controlled by a net control station, and can be used to ensure that connectivity is provided where it is required.

QoS. Typically, repeated TDMA systems support only data. While some capacity could be allocated to provide sufficient data rates for voice, insufficient capacity is available to support all the required voice nets in a brigade-sized force.

Multicast capability. Multicast data is typically supported, allowing some TDMA time slots to be used in the same manner as CNR all-informed nets.

Seamless connectivity. A repeated-TDMA network can support seamless connectivity, providing bridging between the nets is available. Provision of seamless connectivity provides an important mechanism for supporting command and control communications.

Security. Integral, military-grade encryption is typically provided.

EP. Protection against jamming as well as LPD can be provided through the use of CDMA, TDMA, and FDMA. In addition, frequency hopping can be used. Typical hop rates are in excess of 1,000 hops per second.

Minimal mutual interference. The use of CDMA aids in minimization of interference between repeated-TDMA systems and with other parts of the tactical communications system.

Power source. Manpack repeated-TDMA equipment can be run for extended periods on batteries.

Operation in all geographic and climatic conditions. Military communications systems are typically designed to meet the full range of conditions.

Flexibility. Deployment of a repeated-TDMA network is limited by both the maximum area that can be supported on a single net (up to approximately 47 × 47 km for EPLRS) and by the density of stations required to maintain connectivity. Weight can be a limitation for foot-mounted operations, with the weight of the radio set being approximately 12 kg.

7.4.4.2 Current Availability

EPLRS is an example of a repeated-TDMA system currently in service with the U.S. Army. The cost of radio sets is approximately $27,000. A similar capability is also provided by TADIL-J (Link-16), although this system has been designed primarily to support sea and air operations and its terminals are more expensive by approximately a factor of 10.

7.4.5   Base-Station Architectures

The potential solutions for the combat radio subsystem discussed so far have not required the support of any network infrastructure. A range of options exists, however, that make use of a base station in their architecture. These are cellular telephones (or more generally, PCS), trunked-radio systems, and two-frequency, half-duplex radios. Note that we consider satellite-based PCS as overlaid systems, and therefore do not include them as part of an organic tactical communications system.

7.4.5.1   Cellular Telephone/PCS

Cellular telephone/PCS systems provide a circuit-switched communications network. A base station is required for network control, including affiliation of users. All traffic from a mobile station is passed directly to a base station, from where it is routed to its destination. Direct mobile-to-mobile communications is not usually possible. Data rates of up to 9.6 Kbps are possible with current systems such as GSM. Higher data rates, up to approximately 300 Kbps, can be achieved with packet-based extensions of GSM such as GPRS and EDGE and wideband CDMA systems.

The maximum distance between mobile users and the base station is limited to line-of-sight, largely because of the use of UHF and higher frequencies. In a tactical environment, low antennas will significantly reduce communications ranges from base stations. This range could be extended by an elevated base station, which could be carried on an airborne platform or satellite. In the case of TDMA systems, such as GSM, there is a further limitation imposed by the requirement to synchronize all transmitters to a common slot timing. In the case of GSM, this limits the distance between base and mobile stations to approximately 35 km.

Deployment of commercial cellular telephone systems requires extensive path profile planning and measurement to ensure that effective area coverage is achieved. It is doubtful if effective coverage could be achieved by a moving base station, although this may be overcome by a step-up procedure. Restrictions on the location of base stations, especially in forward areas, may also make area coverage very difficult to achieve.

While current technology permits construction of robust, light-weight mobile handsets, base stations are bulky and would be either fixed or vehicle-mounted.

7.4.5.2 Trunked Radio Systems

Digital trunked-radio systems are aimed at a range of voice and data applications that have been met previously by the allocation of dedicated radio channels. Applications include a range of public safety communications systems (including police, ambulance, and fire) and commercial communications systems, such as those used by taxis. The primary mode of operation for trunked-radio systems usually requires that all communications pass through a base station, although some systems support a direct mobile-to-mobile mode. Digital trunked radio systems provide typically a combined voice/data network. A user terminal typically supports a single voice channel and a data capability with capacities similar to or greater than found in typical PCS systems.

The two major digital trunked-radio systems in use are TETRA (an abbreviation for Terrestrial Trunked Radio), which was developed by the European Telecommunications Standards Institute (ETSI) [8], and APCO 25, which was developed by the Association of Public-Safety Communications Officials (APCO) International [9].

7.4.5.3 Two-Frequency Half-Duplex Radio

A two-frequency, half-duplex radio system is illustrated in Figure 7.6. Two radio channels are used. The base station transmits in one of these channels, while all mobile stations transmit in the other. All stations hear transmissions made by the base station. An all-informed net can be created if the base station rebroadcasts signals it receives from other stations on the net.

Other properties of a two-frequency, half-duplex radio system are the same as those for CNR.

7.4.5.4 Military Utility

Range/capacity/mobility trade-off. User terminals typically support a single voice channel and throughputs up to 9.6 Kbps for PCS, 7.2 Kbps for TETRA, and 4 Kbps for a military two-frequency, half-duplex radio. Aggregation of channels may allow higher capacities. In TETRA, for example, four

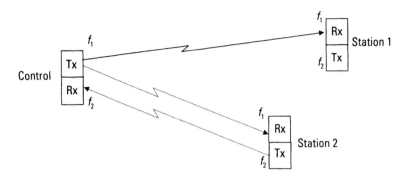

Figure 7.6 A two-frequency, half-duplex radio system.

channels may be aggregated to provide a capacity of 28.8 Kbps. The efficiency of use of the electromagnetic spectrum for all these systems is similar to VHF CNR with a channel bandwidth of 25 kHz. A significant loss in efficiency may result, however, where multicast connections are required if one channel must be allocated for each station participating. While user mobility is greatly assisted by the use of a base station, base-station mobility is problematic for two reasons. First, there is not sufficient time for path planning to be carried out, and it is therefore unlikely that network connectivity will be guaranteed to all subscribers to the same extent that it is in commercial networks. Second, antennas will be mounted on masts to maximize the area of coverage and vehicles will have to remain static while masts are deployed. The ability of TETRA mobile stations to communicate directly without the assistance of a base station greatly enhances the ability to command and control on the move compared to cellular telephony. A base station is still required, however, to provide an interface to external systems such as the tactical trunk subsystem, which limits its mobility.

Command and control on the move. There is no doubt that command and control on the move would be enhanced by the provision of a battlefield-wide cellular network. Commanders would be able to roam throughout the network as dictated by the tactical environment and affiliate to the closest base station. It is doubtful, however, whether sufficient infrastructure could be provided to facilitate such mobility, particularly in forward areas.

Multiple access. Multiple access is achieved by the use of one or more of CDMA, TDMA, and FDMA. Cell-based access offers more opportunity to reuse frequencies than is possible with traditional fixed-frequency CNR nets.

Support for the chain of command. For PCS and trunked radio systems, connections must be established before communications can occur. No automation of this process is supported by off-the-shelf systems. Additionally, in cellular systems, all-informed communications are difficult to provide. While conference calls can be initiated, the number of duplex channels required rises in proportion to the square of the number of stations. Two-frequency, half-duplex systems are able to provide all-informed communications by repeating at the base station.

QoS. QoS for both voice and data is supported.

Multicast capability. Multicast can be achieved in circuit switched systems by the use of conference calls. The major disadvantages of this approach are its poor efficiency of use of the electromagnetic spectrum and the initial complexity of establishing the multicast. Two-frequency, half-duplex radios provide much more efficient multicast.

Flexibility. Flexibility is limited by the requirement for a special-purpose base station. Current base-station implementations are not suitable for man-pack operations. This difficulty in mobility can be overcome by the use of an elevated base station, which would provide a larger coverage area. Additionally, the location of base stations is likely to be severely constrained in support of combat forces—it is unlikely that an unprotected base station would be deployed closer than approximately 20 km from the forward edge of own forces, which renders base station-oriented systems almost unusable for CNR replacement for foot-mounted troops. Further, base stations must be interconnected by high-capacity trunks (probably through the tactical trunk subsystem), which limits the ability of smaller units or subunits deploying without significant infrastructure. For mechanized formations, the ability to protect a base station within an armored vehicle allows deployment closer to combat troops, although high-capacity trunks must still be available to interconnect base stations.

Seamless connectivity. Interfaces to the public switched network exist from PCS and trunked radio systems. Interfaces to military-specific systems could be achieved either directly through specially designed interfaces or indirectly through the strategic network. Two-frequency, half-duplex radios have similar connectivity properties to conventional CNR.

Security. Commercial off-the-shelf systems do not provide military-grade encryption. The use of military-grade encryption on an end-to-end basis is facilitated by the TETRA standard, however, although not necessarily supported by commercially available equipment. Security in military two-frequency, half-duplex systems would typically be provided in the same manner as fixed-frequency CNR.

EP. PCS and trunked radio systems do not provide an EP capability. Even systems such as CDMA employing DSSS do not provide an effective EP capability because, while the spreading codes are effective to provide multiple access, they are not sufficiently secure to provide LPD. PCS also have an unintentional EP capability in that adversary intercept assets would have to acquire the appropriate access channel of a particular base station and then follow the call as it is handed off between base stations.

Minimal mutual interference. PCS and trunked radio handsets and base stations are designed for operation in close proximity to other terminals. Frequency management requirements for a military two-frequency, half-duplex radio are similar to those for conventional CNR.

Power source. Mobile stations can be operated for long periods with very small batteries. Larger power supplies are, however, required for base stations. In GSM, this is accentuated by the requirement for the base station to transmit all the time to provide timing for the network.

Operation in all geographic and climatic conditions. Ruggedization of commercial terminals will most probably be required to enable them to operate across the full range of climatic conditions required, especially with regard to immersion in water and the temperature range.

7.4.5.5 Current Availability

Commercial cellular telephone and PCS systems are available off-the-shelf. They do not, however, incorporate any military security or EP. Commercial systems based on the TETRA standard are available off-the-shelf. They also do not incorporate any military security or EP. A range of commercial and military two-frequency, half-duplex radios is available off-the-shelf.

7.4.6 UWB Radio

UWB radio, also known as impulse radio, transmits information in a sequence of short pulses, typically between 0.1 and 1.5 ns. As illustrated in

Figure 7.7, the information content is encoded in time, rather than frequency or amplitude. In its simplest form, a one may be encoded as a pulse arriving shortly before a nominated time, a zero as the pulse arriving shortly after this time [10].

While a uniform pulse-train spacing may be used (i.e., $T_j = jT_i$), multiple access is best supported by a system incorporating a pseudorandom pulse-train spacing, sometimes referred to as time-hopping. The use of a pseudorandom pulse-train spacing (i.e., the sequence T_j chosen to be pseudorandom) prevents the loss of a large number of consecutive bits due to inadvertent synchronization between two transmitters. The use of a pseudorandom pulse train to provide multiple access for two transmitters is illustrated in Figure 7.8. The variation in the pulse-train timing prevents regular collisions between transmitters, but guarantees that some clashes will occur. Unlike CSMA, collisions that cause loss of data in impulse radio are not primarily due to over-transmission, but arise when the receiver receives two impulses indicating conflicting values for that symbol. The near-far effect prevents the design of a multiple-transmitter, multiple-receiver UWB radio system that uses synchronization of transmitters to overcome collisions.

The use of baseband pulse modulation enables impulse radio to have an extremely wide bandwidth, typically in excess of 25% of the center frequency of the signal. The relationship between pulse length and the frequency content of transmissions is illustrated in Figure 7.9. The transmission of information using time-modulation of short pulses is a form of spread-spectrum communications. Spreading gains of 45 dB (30,000) have been achieved in prototype systems [11]. Like the other forms of spread-spectrum

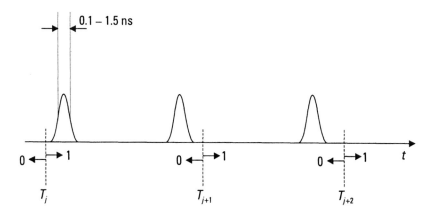

Figure 7.7 Time-coding of information in UWB radio.

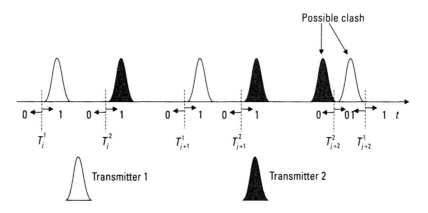

Figure 7.8 Multiple access in impulse radio.

communications discussed in Chapter 5, this frequency spreading can also provide LPD.

The use of baseband transmission creates the possibility of building extremely simple receivers, without the requirement for a conventional RF front end. These receivers have the potential of being much lower cost than traditional receivers based on FDMA.

Very low transmission power levels can be used. Once a receiver is synchronized to a transmission, it does not need to detect pulses. In the example shown in Figure 7.7, the receiver has only to decide whether it is more likely that the pulse arrived before or after the nominated time, enabling the use of low transmission powers for the impulses. The average power is further greatly reduced by the fact that the transmitter is active for only a very small proportion of the time. Average transmitter powers are expected to be on the order of 1 mW. The initial synchronization can be achieved by the transmission of a long synchronization sequence, or by the transmission of a short, higher-power synchronization sequence.

Communications applications for UWB radio include short-range, high-capacity communications systems. It is also possible to construct positioning systems with very high accuracy (better than 1m) and radar imaging applications capable of operating through walls and other obstructions.

7.4.6.1 Military Utility

Range/capacity/mobility trade-off. UWB radios typically provide short-range communications, with capacities either similar to or greater than conventional CNR.

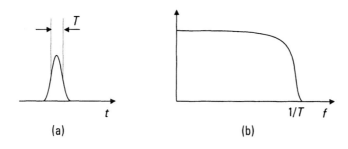

Figure 7.9 Relationship between (a) pulse length, and (b) frequency content.

Command and control on the move. UWB radios support command and control on the move in the same way as conventional CNR, and can provide both voice and data communications.

Multiple access. Multiple access is provided by time hopping. The efficiency of use of the electromagnetic spectrum may be lower than for other forms of multiple access.

Support for the chain of command. UWB radios can support the chain of command by operating with the same net structure as used for CNR.

QoS. QoS is provided by UWB radio in the same way as conventional CNR, allowing efficient support for both real-time and non-real-time services.

Multicast capability. UWB radios are able to provide multicast for both voice and data.

Flexibility. UWB radios are likely to be useable over only short ranges. This is because only low transmit power will be permissible, due to the interference caused to other communications systems.

Seamless connectivity. Interfaces to other nets or other communications subsystems may be provided in the same manner as for any other candidate technology for the combat radio subsystem.

Security.   Security in a UWB network may be through the use of a secure pseudorandom time hopping sequence, or the overlay of conventional encryption on the digital data transmitted.

EP.   A very high level of protection against detection, intercept, and jamming can be obtained if a sufficiently secure pseudorandom time hopping sequence is employed. The requirement to use low transmitter powers to limit interference to other communications systems further reduces the likelihood of detection and intercept, but also limits antijam performance.

Minimal mutual interference.   UWB radios may cause significant interference to a range of other communications systems, especially if they are operated at high power.

Power source.   Because of their very low-power operation, UWB radios are likely to provide long battery life. The potential simplicity of receivers also has the potential to increase battery life beyond that available from more conventional communications systems.

Operation in all geographic and climatic conditions.   Military UWB radios could be built to incorporate military standards or robustness.

7.4.6.2 Current Availability

UWB radio has the potential to provide high capacity, mobile communications within small areas, essentially removing the requirement for frequency management. Multiple access is achieved by assigning a different time-hop sequence to each user community. There remain significant challenges, however, in establishing the feasibility of providing efficient multiple access using impulse radio and in preventing interference with other communications and navigation systems. For example, current systems are believed to cause interference with GPS to a range of approximately 30m [12]. The design of antennas that provide efficient operation with very high bandwidth is also an open problem. Additionally, current regulations for frequency management in most countries explicitly prohibit the use of broadband transmitters that operate across bands allocated for other purposes. These regulations tend to be particularly strict for frequencies allocated for safety and emergency uses. In addition to the technical challenges, therefore, there is a requirement for the reframing of regulations to accommodate UWB radio [13].

7.4.7 Preferred Solution

A summary of the relevant characteristics of the various potential solutions is shown in Table 7.1. Areas of unacceptable performance are shaded in gray.

Data-enabled CNR is unsuitable because it lacks routing between nets and automatic rebroadcast within nets, and is therefore unable to provide seamless connectivity. Repeated-TDMA systems are unsuitable because they do not support voice, do not have sufficient capacity and tend to require high power. Base-station architectures are unsuitable because of the requirement to have a base station that will have limited mobility and that cannot be organic to all the units that it supports.

Preferred combat radio solution. In the above analysis, only the packet radio solution is capable of meeting the full range of basic requirements for communications on the battlefield with fully mobile infrastructure. This solution is therefore preferred as the basic communications system that is organic to all units. An ad hoc network capable of being manually configured to accommodate dispersed operations may be a long-term alternative once this technology matures. The capability of both these solutions will be greatly enhanced by developments in the area of software radio [14].

Limitations. The major shortcoming of packet radio is the low data rates available, which is not sufficient to support some digitization requirements, such as real-time situational awareness. Addressing this issue is the primary purpose of the tactical data distribution subsystem (see Chapters 4 and 8).

7.5 Squad Radio

CNR has traditionally provided communications down to company level, leaving platoon and section commanders to shout orders or use hand signals. One important reason is that the weight of conventional CNR (up to approximately 10 kg, including battery) is too great for a radio to be carried by each soldier. Early U.S. development of a small radio for use at the squad level foundered, however, for the lack of definitive requirements. Initial requirements in the late 1940s for a 9-pound radio were modified in 1949 to seek a 1-pound radio with a 500-yard range, an impossible requirement in the light of the technology available at the time. Portability was also an issue, with the Army finally testing a solution that comprised belt-mounted radios and receivers embedded in fiberglass helmets. In an attempt to minimize weight and cost, the Army decided to develop a production model of the

Table 7.1
Summary of Characteristics of Candidate Solutions

Requirement	Data-Capable CNR	Packet Radio	Ad Hoc Network	Repeated TDMA	Base-Station Architecture	UWB
Range/ capacity/ mobility trade-off	High mobility, low capacity	High mobility, low capacity	High mobility, low to moderate capacity	High mobility, low capacity per user	High mobility, low capacity	High mobility, moderate capacity
Command and control on the move	Good	Good	Good	Moderate	Poor	Good
QoS	Voice and data	Voice and data	Voice and data	Data only	Voice and data	Voice and data
Multicast capability	Good	Good	Good	Good	Often very inefficient	Good
Flexibility	Good	Good	Poor for dispersed operations	Poor for dispersed operations	Poor: Base station is a liability	Poor for dispersed operations
Seamless connectivity	Poor	Good: Data only	Good: Data only	Good: Data only	Requires appliqué	Requires appliqué
Security	Appliqué	In-built	In-built	In-built	None in commercial systems	In-built
EP	In-built or appliqué	In-built	In-built	In-built	None in commercial systems	In-built
Power source	Low power	Low power	Moderate power	Moderate to high power	High power for base station	Low power

squad radio that provided squad leaders with small, handheld transmitters and all squad members with receivers clipped onto their helmets. The PRT-4 transmitter (18 ounces) and PRR-9 receiver (8.6 ounces) were hailed as the answer to the infantry soldier's need to talk in the dense vegetation of South Vietnam. However, to reduce weight, the batteries were strapped unprotected to the radio, and the heat and humidity turned them into masses of dripping cardboard. Not realizing that the helmet served as part of the antenna, soldiers tried to use the receivers apart from the helmet and were disappointed with the reception. Despite test findings that squad members needed only receivers, soldiers in South Vietnam were unhappy without a means to respond to directions. Following a period of heavy use, the squad radio gradually disappeared from the battlefield as soldiers left the radios behind when going out on patrol and commanders consigned them into footlockers for safekeeping [15].

More recently, there has been a resurgence of interest in squad radio in support of programs aimed at developing soldier-level combat systems, such as Land Warrior in the United States, the Future Integrated Soldier Technology (FIST) project in the United Kingdom, and Fantassin à Equipement et Liason Integrées (FELIN) in France.

The basic requirements for squad radio are very similar to those described in Section 7.2. The major differences are:

- Only very short range is required, perhaps 1,000m at the most (British requirements suggest a maximum range of only 500m).
- The requirement for low weight is paramount, which tends to mean low transmit power.
- The ability to carry data and the requirement for seamless integration may be sacrificed, especially for near-term solutions.
- The requirement for encryption to provide security may be sacrificed if LPD is provided by the use of low transmit power from antennas that are placed close to the ground.
- If the radio is small in size and cheap, ruggedization to military standards may not be required, as small size itself provides some physical protection and a replacement of a low-cost radio may be a suitable alternative to repair.
- The need for long battery life is increased by the number of radios (potentially one for each soldier) that must be supplied, making it vital to achieve the goal of 24 hours of operation without replenishment.

An important limitation on the compactness of CNR equipment has always been the requirement that it be possible for an operator wearing NBC gloves to manipulate all controls. This prevents miniaturization of keypads in the way that has occurred for cellular telephones and other commercial products. An alternative to this problem for the squad radio is to make the radio sufficiently simple that once it is configured for a task, the operator is not required to further manipulate any controls.

The potential technical solutions for squad radio are the same as the main component of the combat radio subsystem described in Section 7.4. However, the chosen solution may be modified by the altered requirements listed above.

Voice-only CNR. While voice-only CNR was rejected in Section 7.4 as a solution for the main component of the combat radio subsystem, the unique requirements of the squad radio for light weight and short range allow EP and a measure of security to be achieved by the use of low transmit power and low antennas. The use of low power also aids in maximizing battery life, while the availability of such radios built to commercial standards off-the-shelf reduces cost to the point where disposal is a viable alternative to repair.

Data-enabled CNR. For longer-term solutions in which the ability to exchange data such as GPS location data or images from a weapon sight may be required, data-enabled CNR offers many of the same advantages as voice-only CNR, but with the added data capability. Such a radio is likely to offer voice and data encryption, which has associated with it a large administrative overhead for the generation and distribution of key material. In many countries, there are also additional administrative requirements for stocktaking of cryptographic equipment, and the requirement for higher-level security clearances for soldiers using such equipment, especially if such a radio is seamlessly integrated into a battlefield network. Because of these administrative requirements relating to security, there are important arguments against the seamless integration of a squad radio network with the other parts of the tactical communications system.

Packet radio. While packet radio is able to provide a higher data capability than data-enabled CNR, this comes at the cost of additional weight (especially in batteries) and complexity. The short range over which a squad radio is required to operate make it unlikely that the intranet rebroadcast would have significant value, leaving little additional capability over a data-enabled CNR.

Ad hoc network. While the maturity of ad hoc network technology is not yet sufficient to see it included in fielded systems, the promise of a self-forming network that requires no planning and control has clear attraction. Key outstanding issues for a squad radio based on ad hoc network technology include the desirability of seamless integration with the remainder of the tactical communications system (without which the complexity of ad hoc network technology at this level is of questionable value) and, who will carry the interface if seamless integration is required. An ad hoc network also brings with it the same administrative issues of security discussed previously.

Repeated TDMA. Repeated TDMA systems tend to be data, rather than voice, oriented. Such a system may have merit, however, as a long-term solution for the squad radio, especially if there is a requirement to carry video imagery from a weapon sight. The use of encryption to provide security would create the same administrative issues of security discussed above.

Base-station architecture. The key issue with the use of a base-station architecture as a squad radio is: Who carries the base station? Even if it were feasible to build a light-weight base station, the power requirement for the base station would require a significant extra weight in batteries. A base-station architecture is not likely to be successful in the squad-radio role.

UWB radio. In many ways, UWB radio is ideally suited to the squad radio role. It has the potential to provide an effective low-cost, light-weight, short-range communications system. Its major drawbacks, especially in the near term, are the immaturity of the technology and a lack of understanding of the seriousness of interference with other UWB nets and other parts of the tactical communications system.

A number of armies have recently made decisions to purchase radios to meet these requirements. The U.S. Army has selected a voice-only radio, the Icom F3S, whose characteristics are listed in Table 7.2, for use as a soldier intercom. The radio is intended for use in certain ranger, airborne, air assault, light infantry and mechanized infantry units. The F3S is manufactured to commercial, rather than military, standards. Extensive testing was carried out in a tropical environment because the humidity and heavy foliage represented a probable worst-case scenario for both propagation and equipment failure. As the first major procurement of its Bowman program, the British Army has purchased a similar radio for use as a personal role radio. In both cases, the use of commerical off-the-shelf (COTS) equipment has enabled short lead times in the procurement process. A number of armies have also purchased the

AN/PRC148 Multiband Inter/Intra Team Radio (MBITR), which provides a higher level of performance (including encryption and data), but at the expense of higher weight, cost, and much shorter battery life, as shown in Table 7.3.

7.6 Migration of Analog CNR Systems

Many armies have only just introduced analog CNR subsystems into service at a significant cost. It is not reasonable to expect that the current single channel subsystem can be discarded within the next 10 to 15 years. So, while a packet radio network solution is preferred, procurement of such a system will not solve the short-term problem. Therefore, the most reasonable, cost-effective strategy is to migrate existing CNR systems to be capable of providing a packet radio network.

With regard to the provision of a packet radio network, the primary deficiencies of the in-service analog CNR equipment are the poor support for digital data, including the inability to directly interface to the asynchronous

Table 7.2
Characteristics of Icom F3S Radio

Operating mode	FM voice
Frequency range	136–174 MHz
Channel bandwidth	25 kHz
Channel step	12.5 kHz
Battery life (1:1:8)	24–42 hours (rechargeable)
	21–36 hours (disposable AA cells)
Battery voltage	9.6V
Weight (including battery and antenna)	370g
Transmit power	Up to 5W
Dimensions (including battery)	54 × 132 × 35 mm
COMSEC	Nil
Data interfaces	Nil
Ruggedization	Nil

serial communications port of a computer; the low data rates that can be achieved (with raw data rates no higher than 16 Kbps for VHF and 2.4 Kbps for HF); and the lack of a packet controller that enables either routing of data between nets or intranet rebroadcast within nets.

7.6.1 VHF Radio

Depending on the type and age of the analog radios, the following enhancements may be required to upgrade current systems to a packet radio network: provision of internal or external packet controllers to provide intranet rebroadcast within nets and routing between nets; provision of an interface between the external packet controller and the analog radio; and provision of an interface between the user terminal and the external packet controller.

Options for increasing the throughput of any current data channel should be investigated. Using the current frequency-shift-keyed modulation,

Table 7.3
Characteristics of AN/PRC148 (MBITR)

Operating mode	FM voice, AM voice, digital voice (12 or 16 Kbps)
Frequency range	30–512 MHz
Channel bandwidth	12.5 or 25 kHz
Channel step	5 or 6.25 kHz
Battery life (1:1:8)	8 hours (3 Ah rechargeable lithium ion or lithium disposable AA cells)
Battery voltage	16V (nominal)
Weight (including battery and antenna)	900g
Transmit power	0.1, 0.5, 1, 3 or 5W (FM); 1 or 5W (AM)
Dimensions (including battery)	2.625 inches in width × 7.75 inches in height × 1.50 inches in diameter
COMSEC	Integrated COMSEC (VINSON compatible); frequency hopping upgrade available
Data interfaces	GPS interface
Ruggedization	Immersion to 2m

it is unlikely that a throughput of more than 2 Kbps will be achieved. The architecture of the secure voice system in the radio may constrain the maximum user data rate to 16 Kbps. Significant improvements in the throughput may, however, be obtainable by changing the modulation to provide a higher raw data rate. If this is done, it may be possible to provide internal forward error correction so that the throughput approaches 16 Kbps.

The complexity of the external packet controller will depend heavily on the level of functionality provided. Further work would be required to establish whether it is feasible to provide a full intranet rebroadcast. Improvement of the data-carrying capacity of the radio and its interfaces without this feature may still provide a significant improvement in capability. While consideration should be given to the use of standard protocols such as MIL STD-188-220a and ATM, both complexity and performance may dictate the use of an alternative proprietary protocol between radios with conversion to a standard protocol at the gateway between the radio and user terminals or other communications systems.

It should also be noted that there are significant issues associated with the radiation hazard posed by most current VHF and HF CNR. These issues should be addressed in any modification considerations.

7.6.2   HF Radio

The following enhancements would be required to upgrade current in-service HF radios to a packet radio network: provisions of a high-speed modem, an internal or external packet controller, an interface between the external packet controller and the radio, and an interface between the user terminal and the external packet controller. Depending on the age of the HF radio, it may not be economically viable to upgrade to a packet radio, although much higher data rates would be available from an improved modem.

7.6.3   Resource Implications

It should be noted that it is likely that there will be a significant resource cost associated with migrating the existing CNR variants to packet radios. Funding would be required to address retraining of technicians, drafting training documentation, rewriting of test program sets for any CNR maintenance subsystems, additional purchases of spare parts, investigation and establishment of supporting technical data, and drafting and reproduction of supporting documentation.

7.7 Conclusions and Recommendations

The ongoing requirement for a communications system with fully mobile infrastructure will be met by the combat radio subsystem. Of the available technology, the requirements for this subsystem are best met by packet radio systems, such as those that will soon enter service in the U.S. and British Armies. Other potential solutions lack support for important combat radio subsystem requirements such as voice, security, or EP.

If existing analog CNR have just been introduced into service, procurement of a new radio system may not be able to be justified. Consideration should therefore be given to identifying an upgrade path for in-service equipment to a packet radio network, including technology options, costs, and benefits.

Endnotes

[1] These requirements may sometimes be relaxed in less stressful environments, such as rear areas, allowing less costly radios to be used. The downside of this cost saving in purchase price may, however, be increased maintenance costs due to the increase in the types of equipment that must be supported, which is likely to require a larger inventory of spare parts and consumables (such as batteries).

[2] The advantages and disadvantages of CDMA listed here apply to the use of DSSS as a multiple-access technique. They do not necessarily apply to the use of DSSS for LPI/LPD.

[3] U.S. Army Field Manual FM 24-18, "Tactical Single-Channel Radio Communications Techniques," Sept. 1987.

[4] EPLRS uses TDMA to provide multiple access within an NCS area of responsibility, which is equivalent to a net in the terminology used here. EPLRS supports multiple access within a net by the use of TDMA to provide needlines for communication between stations.

[5] More information on packet radio systems can be found in:

Bertsekas, D., and R. Gallagher, *Data Networks*, 2nd ed., Englewood Cliffs, NJ: Prentice Hall, 1992.

Kahn, R., et al., "Advances in Packet Radio Technology," *Proc. IEEE*, Nov. 1978.

Leiner, B. M., D. L. Nielson, and F. A. Tobagi, (eds.), *Special Issue on Packet Radio Networks, Proc. IEEE*, Jan. 1987.

[6] Rice, M., and A. Sammes, *Command and Control: Support Systems in the Gulf War*, London: Brassey's, 1994, p. 117.

[7]   See, for example:

Johnson, D. B., and D. A. Maltz, "Protocols for Adaptive Wireless and Mobile Net-working," *IEEE Personal Communications*, Vol. 3, No. 1, Feb. 1996.

Perkins, C. E., *Ad Hoc Networking*, Reading, MA: Addison-Wesley, 2001.

[8]   TETRA specifications are contained in:

ETS 300 392-1, "Radio Equipment and Systems (RES); Trans-European Trunked Radio (TETRA); Voice Plus Data (V+D); Part 1: General Network Design," Sophia Antipolis: ETSI, 1996.

ETS 300 392-2, "Radio Equipment and Systems (RES); Trans-European Trunked Radio (TETRA); Voice Plus Data (V+D); Part 2: Air Interface (AI)," Sophia Antipolis: ETSI, 1996.

ETR 300-1, "Radio Equipment and Systems (RES); Trans-European Trunked Radio (TETRA); Voice Plus Data (V+D); Designers' Guide; Part 1: Overview, Technical Description and Radio Aspects," Sophia Antipolis: ETSI, 1997.

ETR 300-2, "Radio Equipment and Systems (RES); Trans-European Trunked Radio (TETRA); Voice Plus Data (V+D); Designers' Guide; Part 2: Traffic Aspects," Sophia Antipolis: ETSI, 1997.

ETR 300-3, "Radio Equipment and Systems (RES); Trans-European Trunked Radio (TETRA); Voice Plus Data (V+D); Designers' Guide; Part 3: Direct Mode Operation (DMO)," Sophia Antipolis: ETSI, 1997.

ETS 300 393-1, "Radio Equipment and Systems (RES); Trans-European Trunked Radio (TETRA); Packet Data Optimized (PDO); Part 1: General Network Design," Sophia Antipolis: ETSI, 1998.

[9]   Standards from APCO Project 25 are published by the Telecommunications Industry Association in their 102-series.

[10] See, for example, Scholtz, R. A., "Multiple Access with Time-Hopping Impulse Modu-lation," in *Proc. MILCOM*, Oct. 1993; or Win, M. Z., and R. A. Scholtz, "Impulse Radio: How It Works," *IEEE Communications Letters*, Vol. 2, No. 1, Jan. 1998, pp. 10–12.

[11] Multiple Access Communications Ltd, "An Investigation into the Potential Impact of Ultra-Wideband Transmission Systems," U.K. Radiocommunications Agency, RA0699/TDOC/99/002, Feb. 2000.

[12] Letter from Office of Spectrum Management, National Telecommunications and Information Administration to the Federal Communications Commission, imposing limits on the grant of a waiver to Part 15 of the FCC rules for UWB radio, June 15, 1999.

[13] A waiver for limited use of UWB radio by fire and police departments in the United States was granted by the FCC on June 29, 1999. This waiver is subject to the use not interfering with other services.

[14] The U.S. Department of Defense is working to develop a software radio system under its JTRS program. See for example: "Joint Tactical Radio System Operational Requirements Document," Washington, D.C.: U.S. Department of Defense, Revision 2.2, Jan. 2001; and "Joint Tactical Radio System Wideband Networking Waveform Functional Description Document," Washington, D.C.: U.S. Department of Defense, Revision 2.2, Aug. 2001.

[15] Bergen, J., *Military Communications, A Test for Technology*, Center for Military History, U.S. Army: Washington, D.C., 1986, pp. 448–450.

8

Tactical Data Distribution Subsystem

8.1 Introduction

The requirement for providing real-time situational awareness creates difficulties for both the combat radio and tactical trunk subsystems. The first of these difficulties is the required communications capacity, which may exceed 500 Kbps for a mechanized brigade, or 2 Mbps for a division. This capacity requirement is made more onerous by the requirement to distribute much of this situational awareness data to a large number of recipients. In some circumstances, a user community for this data may cover the whole of a brigade area. In addition, much of this data must be exchanged between moving vehicles.

The combat radio subsystem, whose throughput is unlikely to exceed 10 Kbps on any one net in the near future, is not able to provide sufficient capacity. The tactical trunk subsystem could be dimensioned to provide sufficient capacity, but is unable to provide the required mobility.

The tactical data distribution subsystem addresses these difficulties and provides a high-capacity, homogeneous, wireless, data network across the brigade or divisional area of operations. The tactical data distribution subsystem is similar in concept to the U.S. Army's concept of the ADDS, which provides medium and high capacity data communications to support situational awareness [1]. The place of the tactical data distribution subsystem in the range/capacity/mobility trade-off is shown in Figure 8.1.

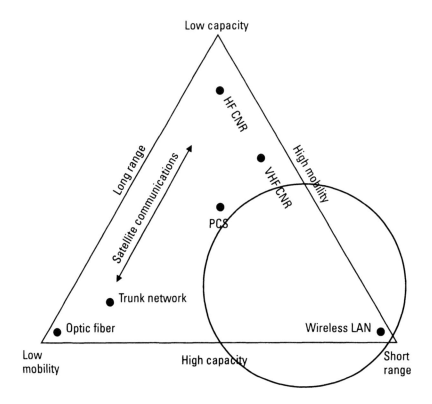

Figure 8.1 Place of tactical data distribution subsystem in the range/capacity/mobility trade-off.

The tactical data distribution subsystem does not need to support voice, which is already well supported at high levels by the tactical trunk subsystem and at low levels by the combat radio subsystem. Indeed, support for voice at low levels would require very high data rates in the tactical data distribution subsystem, possibly in excess of 1 Mbps for a brigade. As is discussed in the following sections, these rates are very difficult to achieve.

While its main purpose is the efficient carriage of data for real-time or near-real-time situational awareness, the tactical data distribution subsystem can also be used to carry other types of data, such as reports and returns. A general characteristic of this traffic is that it tends to consist of a large number of small messages. The provision of situational awareness requires that equipment of the tactical data distribution subsystem are fitted down to the very lowest levels. In a mechanized force, equipment from the tactical data distribution subsystem should be fitted in every armored vehicle. Indeed, a great motivation in the U.S. Army for providing real-time situational awareness is the minimization of fratricide, which requires that the location of all friendly and adversary elements be known with sufficient accuracy to allow them to be distinguished and to allow targeting of the adversary. This equipment should be provided up to at least brigade, going higher (to division or corps) if the tactical trunk subsystem is not able to support the data requirements of real-time or near-real-time situational awareness.

This chapter discusses the basic requirements for the tactical data distribution subsystem and presents characteristics of a number of systems in-service around the world. The chapter concludes with an overview of issues that require further study.

8.2 Key Architectural Drivers

The basic requirements for the tactical data distribution subsystem are as follows.

Range/capacity/mobility trade-off. The minimum capacity of the tactical data distribution subsystem should be sufficient for carrying real-time situational awareness data. As discussed in Chapter 5, it is likely that the minimum requirement is on the order of a few hundreds of kilobits per second, which is shared among a user community consisting of a brigade- or division-sized force. Lower capacities may be acceptable where the force moves only slowly, but higher capacities may be required for dispersed forces, as the requirement for rebroadcast will increase. The total capacity available to the force is likely to be the primary concern in the tactical data distribution subsystem, in contrast to the combat radio subsystem where the primary concern was the capacity available to individual terminals. Given these requirements, the tactical data distribution subsystem is most likely to operate in the UHF or high frequency bands. Operation in the SHF or higher bands is likely to be limited by clutter loss and atmospheric attenuation. The tactical data distribution subsystem also requires high mobility, and therefore must support retransmission to provide communications between terminals that are not within line-of-sight range.

Command and control on the move. Like the combat radio subsystem, the tactical data distribution subsystem must be capable of operation while on the move without stopping. This requires that either there is no ground-based infrastructure or that this infrastructure is fully mobile. The requirements of command and control on the move makes line unsuitable as a sole means of communications; it favors, however, the use of radio with omnidirectional antennas. All radios and terminals must also be sufficiently small and robust enough that they can operate mounted in vehicles. Some terminals should also be capable of operating in a manpack role, although these terminals may have a lower level of capability than those mounted in vehicles.

Multiple access.   The spectrum available for military use is not likely to expand, while the variety of systems that make use of the electromagnetic spectrum increases constantly. Sharing of the electromagnetic spectrum between users is required. Possible multiple access techniques include FDMA, synchronous TDMA, and CDMA. Multiple access may be provided using a combination of these basic techniques. In systems such as EPLRS, multiple access between user communities is achieved using a combination of CDMA, FDMA, and frequency hopping. Multiple access within a user community is provided by TDMA.

Support for the chain of command.   Support for the chain of command requires that communication between a commander and subordinates is achieved with maximum efficiency. Support for the chain of command also requires that ground-based equipment used for communications within a unit or subunit is integral to that unit or subunit. The tactical data distribution subsystem should not constrain the locations of headquarters or other elements. This applies both while on the move and in static locations. A commander should also be able to alter command arrangements within a formation or unit without having to fundamentally restructure the tactical data distribution subsystem. Finally, support for precedence and preemption is required, including a formal management capability.

QoS.   The tactical data distribution subsystem would normally support only data (i.e., non-real-time services) and would normally be optimized to carry real-time situational awareness data. While not imposing the same delay constraints as voice, much of this data (such as location information) is time-critical. Such data is usually not sensitive to loss, however, because it is transmitted at regular intervals.

Multicast capability.   An ability to efficiently carry messages and other data with a large number of destination addresses is required to support real-time situational awareness. This is best achieved by the explicit provision of a multicast capability in the tactical data distribution subsystem.

Flexibility. Operation of the tactical data distribution subsystem across the spectrum of operations should be supported, including mechanized and conventional infantry forces. This is well supported by systems that provide automatic relay of data, but implies a more capable network management function in dispersed operations than in high-intensity operations.

Seamless connectivity. The tactical data distribution subsystem should provide a single homogeneous network across a user community, which will often be a brigade or battalion. Movement of data between the tactical data distribution subsystem and the tactical trunk subsystem and combat radio subsystem should also be supported.

Simplicity. The tactical data distribution subsystem should adapt to changes in network topology due to the movement of nodes without direct operator intervention. What limited human input is required should be restricted to specialist communications staff.

Security. Secure communications should be provided at all levels of the tactical data distribution subsystem. The fact that this subsystem carries information on the locations of friendly forces makes the provision of secure communications vital to the security of the force.

EP. EP is required to provide LPD and resistance to jamming. This might include passive EP, such as terrain shielding or the use of directional antennas. It is also likely that extensive active EP, including DSSS and frequency hopping, will be employed [2]. The importance of EP is greater in the tactical data distribution subsystem than in the combat radio subsystem, because many stations make regular transmissions and would otherwise be extremely vulnerable to detection and intercept. The use of some techniques, such as directional antennas, is made difficult by the mobility requirements of the subsystem.

Minimal mutual interference. Equipments forming the tactical data distribution subsystem should be capable of operation in close proximity with other equipments of the same or similar type, including equipments of the tactical trunk subsystem and the combat radio subsystem.

Power source.   Many types of operation, such as dismounted infantry, will require operation of the tactical data distribution subsystem on battery power for extended periods. As discussed for the combat radio subsystem in Chapter 7, continuous operation of equipment for 24 hours should be able to be achieved without replenishment of batteries.

Operation in all geographic and climatic conditions.   The tactical data distribution subsystem should be capable of being operated under conditions of immersion in water, a temperature range of $-20°C$ to $+70°C$, and withstanding extremes of vibration, shock, pressure, and humidity.

8.3   Multiple Access

In the tactical data distribution subsystem, like the combat radio subsystem, many stations transmit RF energy onto a single channel. Multiple access is required to share the electromagnetic spectrum between these stations.

For the purposes of this discussion, we define a user community as a group of stations between which the tactical data distribution subsystem allows direct communications without passing through a bridge or router, terrain permitting. We do not impose any particular structure of these user communities, that is, they need not follow current doctrine; in the extreme of aggregation, each station in the network may be part of a single user community.

Multiple access poses two key questions: How should multiple access be supported within a user community, and what form of multiple access should be used to permit access to a single channel by multiple user communities? The important differences between the tactical data distribution subsystem and the combat radio subsystem are the size of the user communities and their geographic spread. The traditional hierarchical net structure of the combat radio subsystem is an effective means for supporting command and control. This structure is not effective, however, for the larger user communities associated with situational awareness data.

For each of these issues, the utility of the following multiple access techniques is discussed in Sections 8.3.1 and 8.3.2: synchronous TDMA, CSMA, FDMA, CDMA, frequency hopping, and time hopping.

8.3.1   Multiple Access Within a User Community

Synchronous TDMA.   As long as a central control station is available and flexible allocation of channel capacity is not required, synchronous TDMA can be a very effective means of providing multiple access within a user community. With capacity assigned on demand by a control station, TDMA can also provide a very high level of flexibility and support for precedence and preemption. The major disadvantage of TDMA for the tactical data distribution subsystem is the power consumption resulting from transmissions required to maintain synchronization. With current battery technology, it does not appear to be feasible to use TDMA without at least one vehicle-mounted station.

CSMA. CSMA is the access technique used by current CNR, for both analog and digital voice. By shortening transmissions through digitization, throughput in a CSMA system may be significantly increased. The major remaining drawback of CSMA is that it is not possible to achieve throughput higher than approximately 50% of channel capacity for data.

FDMA. As long as flexible allocation of channel capacity is not required, FDMA can be used for providing multiple access within a user community. The major drawback is the difficulty of monitoring more than one channel at a time, which works against the requirement for multicast communications.

CDMA. Because of the near-far effect, CDMA is not suitable for use as a multiple-access technique except where there is only a single transmitter or a single receiver. This is achieved in mobile telephone networks because all transmissions either emanate from or are destined for a base station. In the tactical data distribution subsystem, it is usually the case that all stations are required to be able to transmit and to receive transmissions from all other stations.

Frequency hopping. The use of frequency hopping to provide multiple access within a user community leads to an inflexible allocation of channel capacity and usually makes monitoring of more than one transmitter infeasible. For this reason, frequency hopping is not usually used for multiple access within user communities.

Time hopping. As stated in Chapter 7, time hopping provides a very flexible allocation of capacity to transmitters. However, the complexity of a receiver that attempts to monitor n transmissions is likely to be n times that of a single channel receiver. In the exchange of situational awareness data within a large user community, this level of complexity may be unacceptably high.

8.3.2   Multiple Access Between User Communities

Synchronous TDMA.   The requirement for a synchronization of timing can make the use of TDMA for providing multiple access between user communities infeasible, as was the case for the combat radio subsystem. Where the user communities are strongly overlapping and where extensive use is made of retransmission (both of which are often required for the exchange of situational awareness data between mobile terminals), it may make sense to separate user communities in the same geographic area using TDMA, while using a different multiple-access technique (such as FDMA) to separate these groups of user communities. This use of TDMA also facilitates demand-assignment of capacity, increasing the flexibility of allocation of communications resources. Use of TDMA as the sole multiple-access technique to separate user communities spread over very large areas is inefficient because large guard intervals are required to compensate for the difference in transmission path lengths.

CSMA.   Efficient operation of CSMA requires a small number of transmitters and that all transceivers can hear all other transmitters. This is rarely the case across a wide area, making CSMA infeasible as a multiple-access technique between user communities.

FDMA.   FDMA is commonly used for providing multiple access between user communities. In fact, it is a useful way of separating user communities.

CDMA.   The near-far effect usually makes the use of CDMA for providing multiple access between user communities infeasible.

Frequency hopping.   Frequency hopping can provide an effective means of multiple access between user communities, as long as deletions due to over-transmissions can be tolerated. These deletions may be acceptable for data if suitable error protection is provided. In this case, however, error-correction data leads to a loss of overall throughput.

Time hopping.   The use of time hopping to provide multiple access between user communities allows for very flexible allocation of capacity and the removal of the need for frequency management, and makes possible simple, digital receivers. In the tactical data distribution subsystem, however, terminals may be required to retransmit signals associated with different user communities. In this application, time hopping has similar disadvantages to those discussed for multiple access within a user community in Section 8.3.1.

8.3.3 Preferred Multiple Access

Because of the large number of stations typically present and transmitting on a user community in the tactical data distribution subsystem, the requirement for each station to monitor transmissions from a number of other stations, and the need for flexible on-demand allocation of communications capacity, TDMA is the preferred multiple-access technique within a user community.

The preferred multiple-access scheme between user communities is FDMA or frequency hopping. Where DSSS transmission is used to provide low probability of intercept (LPI)/LPD, CMDA may be combined with either FDMA or frequency hopping to form a composite multiple-access technique.

8.4 Possible Solutions

In this section, possible solutions for the tactical data distribution subsystem are reviewed. These solutions are data-capable CNR, packet radio, ad hoc networks, repeated TDMA (such as EPLRS and TADIL-J), base-station architectures, UWB radio, and low-capacity TADILs such as TADIL-A, TADIL-B, and TADIL-C.

Many of the systems reviewed here provide capabilities other than communications, which are not discussed.

8.4.1 Data-Capable CNR

Data-capable CNR is able to provide a high-mobility, low-capacity data system, operating over a wide area.

8.4.1.1 Military Utility

Range/capacity/mobility trade-off.   Only low capacity is available, which means that the high data rates caused by aggregation of data as it passes up the command hierarchy cannot be supported. This lack of capacity essentially rules out data-capable CNR as a solution for the tactical data distribution subsystem.

Command and control on the move.   Data-capable VHF CNR is capable of operation on the move in both manpack and vehicle configurations.

Multiple access.   Multiple access is provided using a combination of CSMA and FDMA. The use of CSMA effectively limits the number of stations per net to approximately 10, preventing direct communication among user communities.

Support for the chain of command.   Little support for precedence or preemption is provided. The allocation of capacity between nets is also relatively inflexible.

QoS.   As long as a net is not congested, relatively low delay can be obtained, so long as preambles are not used for synchronization of crypto-equipment. Delays of several seconds may be caused by such preambles, which are commonly used by the current generation of CNR equipment.

Multicast capability.   Data-capable CNR provides an efficient multicast capability within a net; between nets, however, very poor efficiency is provided. This multicast capability is unlikely to be sufficient for the tactical data distribution subsystem, where there is a large number of overlapping user communities and a high proportion of situational awareness data requires multicast transmission.

Flexibility.   Poor support is provided for changing network topology. Provision of this support effectively requires modification of the net structure, the difficulty of which is increased both by frequency planning and key management.

Seamless connectivity.   Automated connectivity is not supported between nets or into other subsystems. This can be partially overcome by the use of bridges between nets.

Simplicity.   The lack of automated rebroadcast means that significantly higher levels of management are required for dispersed operations.

Security.   Encryption may be provided as integral equipment or an appliqué. The use of preambles for synchronization of decryption devices greatly increases the transmission overhead, and results in very low efficiency in providing real-time situational awareness, which consists of a large number of short messages.

EP. Integral forward error correction may be provided. Other EP, such as frequency hopping may be available by the use of an appliqué device.

Minimal mutual interference. Management of mutual interference requires methods similar to those currently in use for CNR.

Power source. Operation for extended periods on battery power may be possible. The requirement for regular transmissions may, however, reduce battery life significantly.

Operation in all geographic and climatic conditions. CNR equipment is specifically manufactured to meet military requirements for ruggedization.

8.4.1.2 Current Availability

Data-capable CNR is currently in service with many western armies, although as part of the combat radio subsystem rather than the tactical data distribution subsystem.

8.4.2 Packet Radio

The operation of a packet radio system is described in Section 7.4.2. Compared to data-capable CNR, packet radio has a distinct advantage by way of the seamless connectivity provided by internet bridging and intranet rebroadcast.

8.4.2.1 Military Utility

Range/capacity/mobility trade-off. While providing very high mobility, a VHF packet radio network is likely to be capable of only low capacities. The use of higher-bandwidth channels in the UHF band could be used to provide higher capacity.

Command and control on the move. Packet radio provides a high level of mobility, and does not require ground-based infrastructure.

Multiple access. Multiple access is provided using a combination of CSMA (within user communities) and FDMA (between user communities). Using these protocols, demand-assignment of capacity is not usually supported. Efficient operation of CSMA protocols is very difficult if more than approximately 10 stations share a common channel. This hinders the formation of the large user communities required to carry situational awareness data.

Support for the chain of command.   Support for precedence and preemption is not provided within a net. A limited capability could be provided by internet bridging.

QoS.   An appropriate QoS for data is provided.

Multicast capability.   Multicast within nets is supported with high efficiency, as long as all stations can receive directly from all other stations. Multicast between nets is inefficient, requiring separate transmission on each net. For large user communities, multinet retransmission would often be required, which is likely to lead to unacceptable performance.

Flexibility.   As long as sufficient density of stations is available to adequately support intranet rebroadcast and internet routing, packet radio supports operations across the spectrum of operations. However, additional engineering and equipment may be required in dispersed operations.

Seamless connectivity.   Seamless connectivity is provided by the combination of internet bridging and intranet rebroadcast. Full automation of this capability is provided by packet radio systems.

Simplicity.   While the use of a fixed net structure is a useful means for minimizing the technical risk of fielding a packet radio, it greatly reduces the ability of the radio to cope with changes in the internet topology, even though changes in the intranet topology are managed transparently.

Security.   Packet radio systems usually have built-in security devices. The use of preambles for synchronization leads to very low efficiency in handling large quantities of small messages, however, as is required for the tactical data distribution subsystem.

EP.   The use of lower transmit power in packet radio than in data-capable CNR provides one means of EP. Additional measures, such as frequency hopping and FEC, are also likely to be available.

Minimal mutual interference.   Mutual interference is managed using the same techniques as are currently used with CNR.

Power source.   The requirement to regularly transmit data for real-time situational awareness makes long battery life a difficult target for manpack operation. This problem is exacerbated by the retransmission requirements of packet radio system.

Operation in all geographic and climatic conditions. A military packet radio system would be constructed to military standards of ruggedization.

8.4.2.2 Current Availability

Packet radio systems are in use (or currently being introduced into service) for combat radio applications in a number of armies. These configurations, however, do not support the high rates of data required for the tactical data distribution subsystem.

8.4.3 Ad Hoc Networks

As discussed in Chapter 7, the self-forming nature of ad hoc networks has a great potential to reduce the management overheads in mobile communications systems. An important example of an emerging system to provide this capability for applications similar to those of the tactical data distribution subsystem is the U.S. Army's Near-Term Digital Radio (NTDR).

8.4.3.1 NTDR

NTDR is an experimental system being developed by the U.S. Army under the Force XXI program to explore the limits of near-term technology and to provide a technical baseline for development of a multiband, multimode digital radio system. Its architecture is based on:

- A mobile packet radio system without the requirement for fixed or permanently positioned base stations;
- A self-forming mobile communications backbone with up to 1,000 stations, which would usually serve a brigade area;
- A reservation-based and receiver-directed protocol, providing efficiency similar to synchronous TDMA while preserving much of the flexibility of uncontrolled CSMA;
- Network management that controls network membership and provides automatic routing of data over the network.

The NTDR can be viewed as an RF system with an embedded router/gateway such as those found in fixed local area networks.

The NTDR transports up to 288 Kbps of user information for each cluster of users, backbone channel, or point-to-point connection. Additional throughput is available depending on the geographic relationship of users employing automatic frequency reuse and availability of additional frequencies. In a typical brigade application the aggregate throughput can be more than 2 Mbps. Message delivery times are claimed to be less than 0.10 second for a single hop. A typical mission thread comprising four hops takes less than 0.4 second.

The NTDR will operate from a single 28-V dc source. Its output power will be dynamically set between 2 mW and 20W. The operating frequency range is 225 to 450 MHz, with a channel bandwidth of 4 MHz. Because each unit is capable of acting as an information relay point, a network of NTDR may extend over a large area. Dynamic multipath equalization is claimed to maintain the on-the-move range at 50 km per hour to within 10% of the static range.

8.4.3.2 Military Utility

Range/capacity/mobility trade-off. An ad hoc network provides a high level of mobility. It is not yet clear whether this technology is able to provide the efficient use of the electromagnetic spectrum that is required to provide high capacity. Given the likely inefficiency of ad hoc network protocols, the difference between the raw capacity and the throughput achieved may be very large.

Command and control on the move. An ad hoc network is ideally placed to provide an operation on the move as long as it is vehicle-mounted. Current ad hoc network technology may support limited-capability manpack terminals.

Multiple access. Multiple access is provided by a combination of synchronous or asynchronous TDMA and FDMA. Flexibility in the allocation of capacity requires some form of assignment of capacity, suggesting that a control station and formal network management function are required.

Support for the chain of command. Support for precedence and preemption requires a network management function. In addition to any automation, some means for human intervention is necessary, suggesting that a completely self-managing network is not necessarily ideal.

QoS. As long as the network does not become congested, an ad hoc network is able to provide an appropriate QoS for data. The difficulties that this technology has with real-time services are probably not relevant to the tactical data distribution subsystem.

Multicast capability.   An ad hoc network based on some form of TDMA is ideally suited to providing a multicast data capability. It is likely to be efficient only within clusters, suggesting that cluster formation should be influenced by the makeup of user communities as well as geographic localization. Once again, this suggests some level of human input to complement automated network management and planning.

Flexibility.   An ad hoc network requires a relatively high density of stations for efficient operation. This may not be achieved in dispersed operations, requiring either a different protocol to be used or a planning function that can position retransmission stations to maintain network connectivity.

Seamless connectivity.   The ad hoc network provides seamless connectivity internally as a natural part of its operation. The provision of suitable interfaces would allow seamless passage of data to other subsystems.

Simplicity.   Ad hoc networks are specifically designed to adapt to changes in network topology. It is not yet clear whether this technology will be able to adapt at the speed of change in topology caused by fast-moving forces, especially in steep terrain.

Security.   Built-in security devices, performing both encryption and key management, are likely to be the norm in ad hoc networks designed for military use.

EP.   A range of electronic-protection techniques is possible. The NTDR radio, for example, provides a combination of DSSS, fast frequency hopping, and narrowband excision at receivers to eliminate effects of narrowband jamming. Further protection against error will be provided by 3/4-rate convolutional coding. A high level of vulnerability may be created by the use of commercial off-the-shelf protocols in the military environment due to their lack of EP.

Minimal mutual interference.   Mutual interference is managed automatically by the ad hoc network, reducing the planning overhead, compared to data-capable CNR and packet radio. It is not yet clear what costs may be incurred in loss of efficiency in the use of the electromagnetic spectrum.

Power source.   Due to the requirement to provide intercluster backbones, at least some stations of an ad hoc network supporting the tactical data distribution subsystem must be vehicle-mounted in order to support high transmission powers.

Operation in all geographic and climatic conditions.   Ad hoc networks designed for military use would be ruggedized to military requirements.

8.4.3.3   Current Availability

No fully functional ad hoc network is yet in service. The implementation of ad hoc networks requires new technology in the formation of clusters and intercluster backbones, the management of network congestion, and the provision of security services, which suggests that this technology still has a high associated risk.

8.4.4   Repeated TDMA

TDMA was identified in Section 8.3 as the preferred means of multiple access within user communities for the tactical data distribution subsystem. Further, TDMA was identified as being an effective means for providing multiple access between user communities where those communities occupy the same geographic area, especially in situations where stations may be a member of more than one user community or move between communities. This section reviews the characteristics of two in-service systems based on TDMA, followed by a general analysis of the military utility of repeated-TDMA systems in the tactical data distribution subsystem.

8.4.4.1   TADIL-J

TADIL-J is a secure, high capacity, jam-resistant, nodeless data link which uses the JTIDS transmission characteristics and the protocols, conventions, and fixed-length message formats defined by the JTIDS Technical Interface Design Plan (TIDP) [3]. TADIL-J operates in the UHF band in the frequency range of 960 to 1,215 MHz, and therefore provides line-of-sight operation. Operation beyond line-of-sight can be achieved by means of a relay, which may be an airborne or satellite-mounted system.

TADIL-J operation is based on all-informed nets. Multiple access between nets is provided by a combination of frequency hopping, FDMA, and CDMA. The 51 channels are supported. Multiple access within a net is provided by TDMA. The TDMA structure is shown in Figure 8.2. *Time slots* of 7.8125 ms are allocated to stations on the net. The 1,536 time slots make up a *time frame*, and 64 time frames form an *epoch*. Each station on the net is allocated at least one time slot per epoch.

Table 8.1 shows the TADIL-J maximum range and data rates, which depend on the operating mode.

EP in TADIL-J is provided by a combination of frequency hopping with an instantaneous hop rate of 77,000 hops per second over 51 frequencies [4], DSSS with a spreading gain of 6.4 and a chip rate of 5 MHz, repeated transmission with data optionally transmitted twice in successive hops, and forward error detection and correction, using a (31, 15) Reed-Solomon code. In each hop, the transmitter is turned on for 6.4 μs, which means that power from a jammer with transmitter-jammer-receiver path length 2 km longer than the transmitter-receiver path length will reach the receiver after the end of the data transmission [5]. TADIL-J contains an embedded cryptographic system to provide secure communications.

In the U.S. Army, TADIL-J terminals will be assigned to division, corps, and *echelons above corps* (EAC). These terminals will support

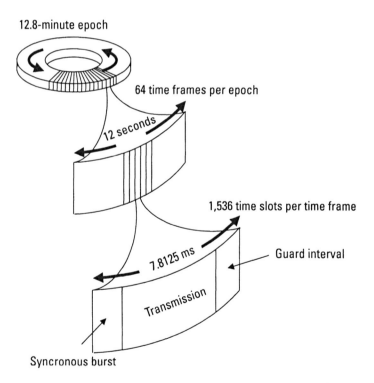

Figure 8.2 TADIL-J time-slot structure.

Table 8.1
TADIL-J Operating Modes

Mode	Guard Interval (ms)	Guard Interval Range Limit (nm)	Throughput After Error Correction (Kbps)	Hops per Second
Standard full slot	4.4585	700	30	33,000
Packed-2 full slot	4.4585	700	59	33,000
Packed-4 full slot	2.0405	300	119	57,000

engagement operations, command and control, surveillance, weapon status and coordination, *precise participant location and identification* (PPLI), and battlefield situation awareness (air and ground).

8.4.4.2   EPLRS

The value of being able to accurately locate friendly forces has long been clear, even though it has not been technically feasible to build an automated position locating and reporting system until relatively recently. Technology developments in the 1970s made possible a radio network in which stations could use time delay to measure their distances from other stations, and with the help of a control facility, to use this information to build up a picture of the locations of surrounding friendly forces. This system, which first saw service with the U.S. Marine Corps, was known as the Position Locating and Reporting System (PLRS). While such a locating system may seem redundant with the availability of GPS, it still provides an important means for locating friendly forces, especially in high-intensity conflicts where the integrity of the GPS system may be threatened locally by jamming or globally by destruction of satellites [6]. The parallel requirement to pass data other than location data led to the development of the EPLRS [7], which is entering service with the U.S. Army.

The two basic types of EPLRS unit are the EPLRS User Unit (EPUU), and the EPLRS NCS. The EPLRS NCS, which is vehicle-mounted, controls the net and provides timing for synchronization. All other EPLRS stations can be either vehicle-mounted or manpacked with a weight of approximately 12 kg. A limited-functionality, light-weight terminal (approximately 3 kg) has also recently been released.

Frequencies in the range of 420 to 450 MHz are used, with the band being segmented into eight channels, each having 3-MHz bandwidth. Multiple access within a net is provided by TDMA technology, in which users transmit information in bursts during predetermined time slots. Multiple access between nets is provided by a combination of FDMA, frequency hopping, and CDMA.

The EPLRS radio supports a variety of data communications services, providing both point-to-point links and an extensive multicast capability, including all-informed nets. Data rates up to 57.6 Kbps per connection, known as a needline, are possible, with rates in excess of 200 Kbps likely to be available in the near future. Common use sees one EPLRS user community per brigade, with a maximum area of approximately 47 × 47 km. Each EPLRS user community has a maximum practical data capacity of between 300 and 450 Kbps, depending on configuration. This capacity is reduced when retransmission is required. Capacity is also reduced when the area over which the user community increases because of the requirement for larger guard intervals between TDMA slots.

The EPLRS NCS has evolved from the PLRS master station through the addition of processing capabilities to automatically manage the initial assignment of the communication paths. The actual relay assignments and evaluation of the performance of the communication paths between host terminals is monitored by the firmware in the individual EPUU. In the event that distribution paths are disrupted due to adversary jamming and terrain, the EPUU will take action to reestablish the required needline. The NCS provides position location, navigation, and identification services and all the network management for the EPLRS control net.

The network management capability also enables EPLRS to automatically build the network from scratch with no prior connectivity information and to automatically adapt to the changing battlefield conditions of terrain masking, user motion, jammer dynamics, and varying subscriber data communications requirements. In addition, the net management design accommodates assignment and deletion of military users to the network.

Continuity of operations in EPLRS is maintained by software that permits data communications to continue along established paths if an NCS is lost. If the loss of a station occurs, the division's NCS or an adjacent brigade's NCS automatically assumes net control of the affected EPUU community. Additional continuity of operations can be gained by the placement of an additional NCS in division rear to assume net control either in planned displacement or during unplanned sudden loss of any NCS. In dispersed operations a brigade may need an internal redundant NCS.

Resistance to jamming is provided through DSSS transmission with a spreading gain of approximately five, frequency hopping operation among eight channels with a hop rate of 512 Hz, error detection and correction, and network management that facilitates the automatic routing and rerouting of messages in the EPLRS network using any EPLRS radio as a relay of opportunity.

Primary power is provided by a 28-V dc battery. The transmit power of the radio is selectable between 0.4, 3, 20, and 100W.

8.4.4.3   Military Utility

Range/capacity/mobility trade-off.   Repeated TDMA systems sacrifice some range for capacity and mobility, but current generation systems are capable of offering several hundred kilobits per second, shared among a large number of terminals. This capacity is appropriate to the provision of situational awareness data for a brigade-sized force. The ability to mount at least some terminals in vehicles is important to ensure that an undue proportion of the network capacity is not devoted to retransmission.

Command and control on the move.   Repeated TDMA systems can be mounted in vehicles to provide full operation on the move. Man-portable systems are less common and tend to offer reduced processing and communications capabilities.

Multiple access.   Multiple access within large users communities (sometimes corresponding to groups of user communities) is provided using TDMA, while user communities are separated by FDMA or frequency hopping. This provides very efficient multiple access, although some control and management by a central control station is required.

Support for the chain of command.   The ability to support large user communities maximizes the flexibility of support for changes in command arrangements. The support of these large user communities is especially important for communications supporting situational awareness.

QoS.   Appropriate QoS for data is supported. Providing low delay voice services may also be possible, assuming that sufficient capacity is available.

Multicast capability.   The use of TDMA as the main multiple-access technique provides a natural support for multicast transmission between stations that are within direct communications range. Efficient rebroadcast can also be supported where range-extension is required.

Flexibility. Support for dispersed operations depends on the provision of effective relays between enclaves. Where enclaves are widely separated, it may be necessary to synchronize separately within enclaves, with timing translation provided by the relay. The management associated with this relay function is naturally performed by the network control station.

Seamless connectivity. Repeated TDMA systems typically provide seamless connectivity internally; the provision of interfaces allows seamless transfer of data with other subsystems.

Simplicity. Rapid changes in network topology may be supported by the provision of relays. Systems designed for use in the land environment (e.g., EPLRS) tend to provide this automatically, although a degree of manual network planning is required to ensure that relays are appropriately positioned.

Security. Built-in encryption is provided.

EP. Built-in EP, including spread spectrum and FEC, is commonly available. In TDMA systems, where most stations are transmitting at regular intervals, high-performance EP systems are of great importance.

Minimal mutual interference. The use of TDMA to provide multiple access alongside spread spectrum techniques minimizes the planning overheads to avoid interference.

Power source. Regular transmission, whether broadcasting situational awareness data or providing retransmission, is likely to lead to short battery life in manpack systems. Although vehicle mounted systems may use transmitter powers up to 100W, manpack systems tend to operate at lower transmitter powers to avoid unnecessarily reducing battery life.

Operation in all geographic and climatic conditions. Being built to military specifications, ruggedization is common in repeated TDMA systems.

8.4.4.4 Current Availability
Both EPLRS and TADIL-J are currently in service with U.S. forces.

8.4.5   Base-Station Architectures

Three base-station architectures were described in Chapter 7. Commercial PCS systems (Section 7.4.5.1), trunked radio (Section 7.4.5.2), and two-frequency, half-duplex radios (Section 7.4.5.3) all provide communications services that could be used in the tactical data distribution subsystem.

8.4.5.1   Military Utility

Range/capacity/mobility trade-off.   The requirement for base stations, which usually require antennas elevated on masts, limits the mobility of base-station architectures. The relatively small number of base stations provides a much lower level of retransmission performance than systems in which any station can perform this task. Even if base stations use low, omnidirectional antennas, some form of semimobile trunking infrastructure between base stations is required, violating the mobility requirements of this subsystem.

Command and control on the move.   Base stations typically require extensive engineering in addition to being stationary during operation, and are unlikely to be able to operate on the move.

Multiple access.   Multiple access is accomplished using FDMA or a combination of FDMA and TDMA. This multiple access tends to be efficient for point-to-point communications, but inefficient for point-to-multipoint communications, which is the primary mode for situational awareness communications.

Support for the chain of command.   Provision of base-station architectures that are integral to all levels of a force is usually not feasible, which would limit the ability of elements of the force to be dispersed on separate tasks.

QoS.   Appropriate QoS is provided for data.

Multicast capability.   Multicast usually requires allocation of capacity for each receiver in cellular telephone and trunked radio systems, reducing efficiency to unacceptable levels. Given that multicast is the primary form of situational awareness communications, this loss of efficiency is unacceptable. Even for two-frequency, half-duplex radio systems, multicast is only efficient for stations that are within the direct communications coverage of a transmitter. Such systems do not usually provide an efficient relay for stations that are outside this range.

Flexibility. As long as sufficient base stations are deployed and sufficient resources are available for inter-base-station communications, base-station architectures can support dispersed forces. The inter-base-station trunk infrastructure would therefore require augmentation for dispersed operations. The ability of a base-station architecture to provide area coverage in dispersed operations may also be limited by the coverage of the limited number of available base stations. This difficulty could be reduced by the use of a base station mounted on an airborne platform, as described in Chapter 9.

Seamless connectivity. Through suitable interfaces, data can be exchanged with other subsystems.

Simplicity. The handoff process supports slow changes in network topology. Fast changes in network topology may overload the processing or signaling capacities of the system. Changes in the topology of the network between the base stations are also usually not supported.

Security. Commercial systems do not usually have military-grade encryption built in. This is currently being addressed by efforts such as the U.S. National Security Agency's CONDOR program, which is working to embed military-grade encryption into a variety of commercial cellular telephone terminals. The cost effectiveness of this program relies on a range of nonmilitary government uses anticipated for these systems.

EP. EP is not usually provided in these systems. Unlike encryption systems, the cost of adding military grade EP to handsets and base stations is likely to be prohibitive.

Minimal mutual interference. Frequency planning, following similar rules to those currently used for CNR, is required.

Power source. Base stations typically require a generator for operation; other stations may be able to operate for extended periods using batteries.

Operation in all geographic and climatic conditions. Commercial systems may require ruggedization. The economic feasibility of this is unclear. Physical ruggedization and construction to support troops operating in an NBC environment are likely to be feasible by modifying the external casing. Electromagnetic ruggedization is likely to involve modification of the internal electronics of handsets and base stations, and may not be economically viable.

8.4.5.2   Current Availability

A range of commercial networks based on a base-station architecture are in common use, including PCS, trunked radio, and two-frequency, half-duplex radio, such as taxi dispatch systems.

8.4.6   UWB Radio

The operation of UWB radio is described in Section 7.4.6. Its primary distinguishing feature is the use of extremely wide bandwidths, with time hopping used to provide multiple access.

8.4.6.1   Military Utility

Range/capacity/mobility trade-off.   UWB radios can potentially provide high mobility and high capacity over short ranges. Provision of coverage over large areas would require the incorporation of relays, probably based on more conventional communications technology.

Command and control on the move.   Full operation on the move is possible.

Multiple access.   Multiple access within and between user communities is provided by time hopping, which may provide less efficient use of the electromagnetic spectrum than other multiple-access techniques, which can be seen as part of the cost of its high level of EP.

Support for the chain of command.   With the use of a control station, the time-hopping multiple access does allow for flexible allocation of capacity and support for precedence and preemption.

QoS.   UWB radio supports both real-time and non-real-time services.

Multicast capability.   Multicast transmission is supported; reception of multiple simultaneous transmissions effectively requires one receiver per received signal, which is likely to increase power consumption in UWB receivers.

Flexibility.   UWB radio may require bridging between enclaves in dispersed operations, which is likely to be based on more conventional communications technology.

Seamless connectivity. A UWB network could provide seamless connectivity following the approach of either packet radio or ad hoc networks.

Simplicity. Dispersed operations are not supported by UWB radio. Some form of bridging (provided by a more conventional radio) between enclaves would be required in such operations, which is likely to be based on more conventional communications technology, and require a high degree of network planning and management by specialist communications staff.

Security. Encryption can be used in conjunction with time hopping. The use of preambles for synchronization leads to very low throughput in the tactical data distribution subsystem. The use of low power to minimize interference with other systems also provides LPD.

EP. The use of time hopping for multiple access is itself a means of EP, which provides a high level of protection against detection and intercept, and a moderate level of protection against jamming (which is limited by the need to use low transmitter powers to minimize interference with other parts of the tactical communications system). FEC may also be used.

Minimal mutual interference. Interference with other communications systems and navigation systems such as GPS is likely to occur.

Power source. Because of their low transmit power, UWB radios are likely to provide high efficiency and a long battery life.

Operation in all geographic and climatic conditions. A military UWB radio would be manufactured to military standards of ruggedization.

8.4.6.2 Current Availability

Experimental UWB radios are in use. Further technical development is required before a UWB network is achievable, such as would be required for use in the tactical data distribution subsystem. Modification of procedures for deconfliction of different uses of the electromagnetic spectrum, which are currently based on the allocation of a band of frequencies to each use, would also be required.

8.4.7 Low-Capacity TADIL

TADILs have provided a highly effective means for sharing situational awareness data in the sea and air environments for a number of years. The characteristics of three low-capacity TADILs—TADIL-A, TADIL-B, and TADIL-C— are examined here, while TADIL-J was discussed in Section 8.4.4.

8.4.7.1 TADIL-A

TADIL-A (also known as Link-11A) employs netted communications techniques using standard message formats. TADIL-A radios can operate in the HF band, giving a range of up to 300 nm, or the UHF band, giving a range of approximately 25 nm surface-to-surface or up to 150 nm surface-to-air. TADIL-A data links operate at rates of 1,364 bps (HF/UHF) or 2,250 bps (UHF). The modulation scheme is differential QPSK [8].

TADIL-A normally operates on a polling system with a net control station polling each participant in turn for his or her data. In addition to this roll-call mode, TADIL-A may be operated in broadcast modes in which a single data transmission or a series of single transmissions is made by one participant. TADIL-A is, therefore, a half-duplex link. TADIL-A is secure but does not provide any specific EP capability.

TADIL-A is used commonly in the sea environment for the exchange of air, surface, and subsurface tracks, EW data, and limited command data among connected terminals, but it does not support aircraft control or other warfare areas.

8.4.7.2 TADIL-B

TADIL-B (also known as Link-11B) provides a secure, full-duplex, point-to-point digital data link utilizing serial transmission frame characteristics and standard message formats at 2,400, 1,200, or 600 bps. It interconnects tactical air defense and air control units.

8.4.7.3 TADIL-C

TADIL-C (also known as Link-4) is an insecure, time-division digital data link utilizing serial transmission characteristics and standard message formats at 5,000 bps from a controlling unit to controlled aircraft. Information exchange can be one way (controlling unit to controlled aircraft) or two way.

8.4.7.4 Military Utility

Range/capacity/mobility trade-off. None of these systems, the most capable of which operates at 5 Kbps, is able to support the data requirements of the tactical data distribution subsystem.

Command and control on the move. The use of HF in some systems limits the capability for communications on the move. Other systems, operating the VHF and UHF bands, do have sufficient mobility.

Multiple access. Multiple access within user communities is usually based on polling by a control station. This increases the vulnerability of the network, unless alternate control stations are available. Multiple access between user communities is provided by FDMA, which provides a high efficiency of use of the electromagnetic spectrum.

Support for the chain of command. Little flexibility is provided in the structure of user communities, primarily due to the low capacity and the requirement for a control station on each net. Flexibility could be increased by providing rebroadcast between nets, but the low capacities of these systems is likely to preclude this in many situations.

QoS. Non-real-time services are supported. Some FEC is provided, but the primary means of error recovery is to wait for the same data to be transmitted again. This works well for data such as platform tracks, where the loss of a single transmission has little impact on the overall track plotted.

Multicast capability. Some systems support multicast data within nets. The low capacities of these systems and the difficulty of providing rebroadcast between nets make the establishment of flexible user communities, as required for land situational awareness, difficult.

Flexibility. Systems such as TADIL-A that are capable of operating in the HF band may provide very effective support to dispersed operations, albeit at only low capacities.

Seamless connectivity. Connection to other subsystems via an interface is possible. Exchange of data between user communities via internal interfaces may also be achievable.

Simplicity. No direct support for retransmission is provided to cope with changes in network topology. A high degree of network management and planning by specialist communications staff is likely to be required.

Security. TADIL-A and TADIL-B provide secure operation; other systems do not. Modification of other low-capacity TADIL equipment to support security is unlikely to be cost-effective.

EP. No active EP is provided.

Minimal mutual interference. Careful arrangement of antennas on a platform and frequency planning are required to ensure that interference is not caused to other systems.

Power source. These systems usually operate at high powers (100W or more) and are therefore not suitable for extended battery operation.

Operation in all geographic and climatic conditions. Being specifically designed for military use, TADILs are constructed with high levels of ruggedization.

8.4.7.5   Current Availability

TADIL-A, TADIL-B, and TADIL-C are all in service with U.S. and allied forces.

8.5   Preferred Solution

A summary of the relevant characteristics of the various potential solutions is shown in Table 8.2. Areas of unacceptable performance are shaded in gray.

Based on the analysis of currently or soon-to-be available systems, the preferred solution for the tactical data distribution subsystem is one based on a repeated-TDMA architecture. This system supports the basic requirements of the tactical data distribution subsystem, the keys to which are command and control on the move, multicast capability, EP, and capacity.

The use of an ad hoc network may provide a higher level of flexibility. Given the current lack of fielded systems, there is a much higher technical risk in the adoption of such a system.

8.6   Conclusion

In most armies, there is no in-service system that attempts to meet the requirements of the tactical data distribution subsystem. Development of this subsystem (including equipment acquisition and the generation of

Table 8.2
Summary of Characteristics of Candidate Solutions

Requirement	Data-Capable CNR	Packet Radio	Ad Hoc Network	Repeated TDMA	Base-Station Architecture	UWB	Low-Capacity TADIL
Range/capacity/ mobility trade-off	Insufficient capacity	Insufficient capacity	Insufficient capacity	High capacity and mobility	Base station must be stationary	Potential for high capacity and mobility	Insufficient capacity
Command and control on the move	Yes	Yes	Yes	Yes	Base station must be stationary	Yes	Yes
Support for chain of command	Poor performance when congested	Poor performance when congested	Possibly poor performance when congested	Yes	Base stations unlikely to be integral assets	Yes	Unlikely to be integral asset at all levels
Multicast capability	Efficient within net	Efficient within net	May not be efficient	Very efficient	Inefficient	Efficient over short ranges	Efficient within coverage of control station
Seamless connectivity	Poor	Good	Good	Good	Good	Good	Good
Security	No	Yes	Yes	Yes	Not in commercial systems	Yes	Yes
EP	Very limited	Yes	Yes	Yes	No	Yes	Very limited

doctrine, procedures, and training) should be a priority if real-time situational awareness is required. This process would begin with a study to ascertain the types of data to be carried, the regularity with which this data must be retransmitted (which may vary greatly depending on the operational scenario and its phase), the data rates required, the network entry and exit points (interfaces) for data, the technology available for meeting these requirements, the operational impacts of introducing this technology, and the subsystem's support requirements.

Based on current technology, solutions based on a repeated TDMA architecture are likely to provide the best trade-off between performance and technical risk.

Endnotes

[1] U.S. Army Field Manual FM 11-75, "Battlefield Information Services (BIS)," Sept. 1994.

[2] See, for example, Chapter 3 of Frater, M. R., and M. Ryan, *Electronic Warfare for the Digitized Battlefield*, Norwood, MA: Artech House, 2001.

[3] See, for example:

"JTIDS/TIES Consolidate Tactical Communications," *EW*, Sept./Oct. 1977.

MIL-STD-6016, "DoD Interface Standard Tactical Digital Interface Link (TADIL) J Message Standard," Feb. 1997.

Stiglitz, M., "The Joint Tactical Information Distribution System," *Microwave Journal*, Oct. 1987.

Toone, J., and S. Titmas, "Introduction to JTIDS," Signal, Aug. 1987, pp. 55–59.

[4] Because guard intervals are required to accommodate the changing distances between transmitters and receivers, the transmitter is turned off for some of the time. The number of hops actually occurring in 1 second is therefore less than the instantaneous hop rate.

[5] Frater, M. R., and M. Ryan, *Electronic Warfare for the Digitized Battlefield*, Norwood, MA: Artech House, 2001, pp. 79–84.

[6] See, for example:

Bond, L., "Global Positioning Sense," Proc. 25th Annual Convention and Technical Symposium, International Loran Association, San Diego, CA, Nov. 1996.

Carroll, J., et al., "Vulnerability Assessment of the U.S. Transportation Infrastructure That Relies on GPS," ION National Technical Meeting, Long Beach, CA, Jan. 2001.

[7]   U.S. Army Field Manual FM 24-41 "Tactics, Techniques, and Procedures for the Enhanced Position Location Reporting System (EPLRS)," Final Draft, July 1999.

[8]   MIL-STD-188-203-1A, "Interoperability and Performance Standards for Tactical Digital Information Link, (TADIL) A," Jan. 1988.

9

Tactical Airborne Subsystem

9.1 Introduction

An airborne communications platform can be used to provide additional capacity and to extend ranges of the combat radio and tactical trunk subsystems when dictated by the operational scenario. Traditionally, solutions to the provision of high-capacity, long-range communications have relied on the use of satellite-based services. However, subspace platforms (i.e., airborne, rather than space-borne) offer a viable alternative, with the potential to deliver a broader range of services more cost-effectively. These platforms are also known as *stratospheric communications platforms*, and are called *high-altitude platform stations* (HAP) by the ITU. Other names include *high-altitude aeronautical platforms* (HAAP) and *high-altitude long-endurance* (HALE) platforms.

The concept of airborne communications is not new—U.S. operations in South Vietnam highlighted the importance of airborne relay in supporting wide-ranging campaigns. For example, during Operation Silver Bullet (October 23–November 20, 1965), the 13th Signal Battalion of the 1st Cavalry Division employed a fixed-wing aircraft operating at 10,000 feet and equipped with 12 FM radios to retransmit voice command nets [1]. While helicopter-borne relay platforms were also developed, the cost of aircraft operation meant that airborne retransmission sets were generally only used as temporary relays while units were setting up ground relays, or during major operations [2]. Additionally, since almost every brigade and battalion commander controlled critical operations from a helicopter-borne command post, there was less need for airborne relays when commanders were airborne [3].

This chapter examines the utility of a tactical airborne subsystem in support of future land warfare. Advantages of an airborne communications platform are discussed and payload requirements are analyzed. Some limitations are examined before an upper-level architecture is presented. Potential platform types are discussed and relative costs are outlined.

9.2 Advantages of an Airborne Communications Platform

Before addressing the issues in detail, it is worth briefly considering the number of significant advantages to the operation of an airborne communications platform, which can offer a broad array of services with low operating costs compared with satellite-based systems:

- Airborne platforms do not require a launch vehicle; they can move under their own power to change position and to remain on-station.

- Platforms can be retrieved and relaunched, allowing the commu -nications payload to be brought down to Earth for routine maintenance.

- Airborne platforms can be steadily enhanced with emergent technologies (unlike satellite communications payloads, which remain fixed in the deployment configuration).

- Once a platform is in position, it can immediately begin delivering services to its service area without the need to deploy a global infrastructure or a constellation of platforms to operate (providing some form of central terminal is available in theater). Coverage can therefore be provided more rapidly to areas of operation without having to have significant satellite assets covering the area of potential tasks.

- The aircraft is 20 to 2,000 times closer to the user than a satellite, with 10 times the available electrical power. The relatively low cost of the platform and gateway stations make it the cheapest wireless infrastructure per subscriber over large areas.

- The low altitudes provide short paths through the atmosphere, reducing attenuation.

- Higher antenna elevation angles provide most subscribers with unobstructed line-of-sight to the platform.

- Airborne platforms have much lower transmission delays—tens of microseconds compared to tens to hundreds of milliseconds for satellite. Lower delays reduce problems with voice communications as well as with data protocols such as TCP/IP.

- With small antennas and low power requirements, an airborne communications platform can support a wide variety of fixed and portable user terminals to meet almost any service need.

- Unlike satellite systems, which are all-or-nothing, multibillion-dollar investments, airborne platforms can be procured and deployed one at a time as strategic guidance and budgets allow.

- Much lower capital investments are required—tens to hundreds of millions of dollars compared to several billion.

The advantages of stratospheric platforms have created considerable interest for commercial communications applications, and a number of subspace systems have been proposed (a number of major systems are discussed in Section 9.7). For example, for PCSs, a metropolitan area can be served within a 100-km radius by one beam of a GEO satellite, by not more than six to nine beams of a LEO satellite, and by as many as 700 to 1,000 beams formed by a stratospheric platform. For stratospheric platforms, the coefficient of frequency reuse is therefore two orders of magnitude higher than satellite-based systems [4].

While these advantages offer significant tactical benefits, it should be noted that, as with any practical system, there are some operational limitations, which are discussed in Section 9.5.

9.3 Platform Height

Although most commercial systems are proposed for operation in the stratosphere, an airborne platform could potentially operate at any height. Table 9.1 shows some representative operating heights that offer possible solutions to requirements for the tactical airborne subsystem.

As can be seen from the table, the lower the altitude, the shorter the communications range to ground terminals, and more platforms would therefore be required to cover a particular area of operations (AO). Additionally, a larger number of platforms would be required at the lower height to maintain 24-hour coverage due to the shorter endurance.

Table 9.1
Possible Platform Heights for the Tactical Airborne Subsystem

Loiter Altitude	65,000 feet (above commercial airspace)	45,000 feet (above commercial airspace)	25,000 feet (in commercial airspace)	13,000 feet (in commercial airspace)
Endurance	42 hours	8+ hours	40+ hours	3 hours
Range	5,500 km/24 hours/5,500 km	900 km/8 hours/900 km	900 km	185–230 km
Footprint Diameter	500–650 km	400 km	300 km	150 km
Payload	900 kg	350 kg	200 km	35 kg

From: [5]

The larger (65,000-feet operating height) platforms such as the U.S. Global Hawk are also self-deployable worldwide, whereas the others are not self-deployable and would require considerable in-theater infrastructure to maintain operations. In particular, a suitable airstrip would be required, although many of the smaller tactical systems can be launched and recovered from mobile infrastructure (e.g., a small UAV can be launched from a trailer and recovered by parachute). These smaller assets would therefore be deployed by uniformed personnel, as opposed to Global Hawk, which could provide a tactical service but have the logistical advantages of operating from civilian airports in the strategic environment, and be repaired and maintained by civilian personnel.

Additionally, an asset that can perform as the tactical airborne subsystem will require the larger payload capacity of Global Hawk. The services detailed in Section 9.2 could not be provided by the smaller platforms, which would only be able to provide range extension for a few combat radio channels.

Larger platforms operating at greater altitudes (approximately 65,000 feet) are therefore preferred as the basis of the tactical airborne subsystem. The deployment of a larger number of less capable systems is not likely to be cost-effective. This is not to say that the less capable systems are not useful at all as a communications relay. Tactical, low-level platforms are potentially very useful to tactical commanders for both communications and surveillance. However, these will normally be organic assets to provide intimate support to tactical units and will not be capable of supporting communications across the deployed force.

9.4 Payload Requirements

The tactical airborne subsystem provides an airborne platform that carries robust communications packages to support command and control across wide areas. High-gain antennas coupled with the ability to loiter at high altitudes for extended periods will enable tactical users equipped with lightweight omnidirectional antennas and low-power radios to establish long-range communications from mobile platforms. This capability will provide a significant improvement in communications ranges and will enhance the ability for commanders to command and control on the move. The tactical airborne subsystem is required to provide range extension of the combat radio, tactical trunk, and tactical data distribution subsystems; and to provide additional communications services, including surrogate satellite communications and coverage extension of overlaid communications systems, such as PCSs and the theater broadcast system (TBS).

9.4.1 Range Extension

The major requirement for a tactical airborne subsystem is to extend the range (mainly through retransmission) of the other subsystems of the tactical communications system where the provision of high-capacity, long-range communications tends to be intractable, particularly to mobile assets.

As illustrated in Figure 9.1, the range extension offered by an airborne communications platform is very significant. In their normal terrestrial modes, CNR and radio relay are generally terrain-limited, not power-limited. For example, ground ranges of CNR are limited to 5 to 15 km depending on how high the operator can elevate the antenna. An airborne communications platform has the potential to extend those ranges to up to 500 km—a dramatic improvement. For example, had an airborne communications platform been available during the Gulf War, true communications on the move could have been provided to support a fast-moving, wide-ranging envelopment, at a time when terrestrial networks were stretched to breaking point [6]. An airborne communications platform offers the opportunity to provide commanders with command and control on the move over a large operational area.

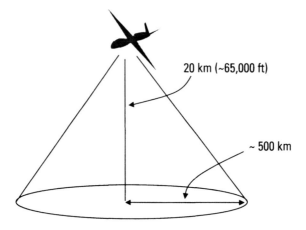

20 km (~65,000 ft)

~ 500 km

Figure 9.1 Footprint of airborne communications platform.

A further extension in coverage can be provided by cross-linking between platforms, as illustrated in Figure 9.2. However, it should be noted that cross-linking requires antennas with fairly high gains that have to be pointed accurately to another airborne platform. Therefore, cross-linking represents some engineering challenges for most platforms except those that can be maintained in a geostationary position, although modern phased-array antennas go some considerable way to solving this problem. The inclusion of a cross-linking capability will reduce the communications payloads onboard the platform.

9.4.2 Additional Communications Services

Once the requirement for a tactical airborne subsystem has been accepted on the basis of range extension for the other subsystems in the tactical communications system, a number of additional communications services can be provided from the airborne platform.

Intrabattlespace communications. As mentioned in Section 9.4.1, an airborne platform presents the opportunity to provide true joint and combined communications in a simple manner. Figure 9.3 illustrates how the airborne platform can relay communications and establish a net between joint assets as well as provide reach-back communications by satellites in either LEO or GEO. The platform could also provide Link-16 relay capability to extend connectivity between widely dispersed air-defense assets, both airborne and land-based. It should be noted, however, that intrabattlespace communica-

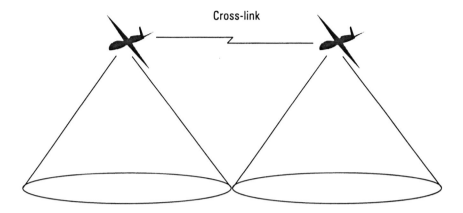

Figure 9.2 Increase in coverage by cross-linking between airborne platforms.

tions are only facilitated by an airborne platform if the various communications systems are themselves interoperable. The interoperability of joint and combined communications systems is a thorny issue that takes considerable, sustained effort to achieve and maintain.

Surrogate satellite communications. It may also be possible to mount a surrogate satellite transponder onboard the airborne platform. In-service satellite ground terminals could be used to communicate to the surrogate transponder rather than to a satellite. Since the airborne platform is much closer, lower powers (higher data rates) are possible throughout the AO, without having to be within the satellite footprint.

Link-16 repeater. Provision of Link-16 (TADIL-J) communications to terrestrial terminals generally poses an intractable problem due to the inability of the terminals to "see" each other. An airborne Link-16 repeater could extend coverage to terrestrial terminals across the AO, greatly facilitating sensor-to-shooter communications.

PCS access. It may be possible to mount a cellular PCS base station on the airborne platform and provide digital mobile telephony coverage within the footprint. However, commercial PCS base stations tend to be large and heavy, although there are some initiatives within the commercial industry to reduce size and weight. There are a number of programs in the United States and Europe that are examining the possibility of airborne PCS base stations.

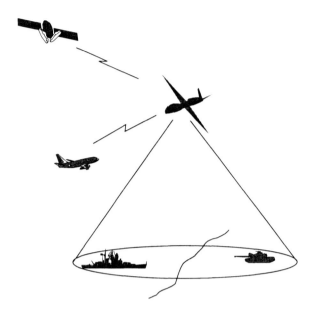

Figure 9.3 Improved battlespace communications.

The large power requirements of a base station provide an additional problem. However, recent studies have shown that a scaled-down base station that accommodates 60 simultaneous calls can be included as part of the U.S. Airborne Communications Node (ACN) concept (discussed in more detail in Section 9.7.2). Even if a base station is not included in the platform payload, the airborne platform has the potential to provide a reach-back capability with digital cellular phones to support connection back to a base station on the ground.

Theater broadcast.   The airborne platform offers two opportunities with regard to theater broadcast: broadcast relay and in-theater broadcast.

- *Broadcast relay.*   One of the difficulties in getting satellite theater-broadcast to tactical units is the difficulty that combat troops have in deploying reasonably sized satellite antennas. Broadcast relay would take the satellite broadcast on the airborne platform and rebroadcast it into the footprint in the VHF and UHF bands. Broadcast communications can then be provided to tactical users who can receive broadcasts with small omnidirectional antennas and low-powered receivers.

- *In-theater broadcast.* It is doubtful whether the brigade commander would allow theater broadcast into the brigade area without modification by brigade staff; the most useful product for brigade units is a broadcast of the brigade/divisional view, rather than a strategic one. In that case, the airborne platform allows the brigade commander to take the satellite broadcast at brigade headquarters, modify and add information, and then provide an in-theater VHF/UHF broadcast through the airborne platform. Again, any mobile units in the footprint can receive high-capacity broadcast into small, omnidirectional (whip) antennas with low-powered receivers.

Videoconferencing. Should commanders decide that videoconferencing is required, terrestrial solutions are difficult to arrange. However, an airborne platform provides an ideal vehicle for the delivery of videoconferencing facilities into the footprint.

Special-forces communications. High-capacity, long-range communications are normally difficult for special forces (SF), which are commonly forced to operate using low-capacity HF to provide adequate range. This significantly reduces the ability of SF patrols to transmit high data rates to support applications such as image transfer for surveillance roles. An airborne communications platform can extend high-capacity communications to SF patrols, which can be equipped with small, low-powered receivers with small antennas that are not required to be pointed accurately.

9.4.3 Potential Noncommunications Uses

In addition to communications, the tactical airborne subsystem may also potentially be used to support EW and remote sensing. Just as the airborne platform provides an ideal communications base, it also provides an excellent platform for EW. Ground-based intercept is invariably terrain-limited, and many more ground assets would be required to have the same coverage as an airborne EW platform. The airborne platform could also provide broad area surveillance of the AO in a range of frequencies, through its ability to carry a range of optical, infrared, multispectral, and synthetic aperture radar (SAR) sensors. In addition to carrying sensors, airborne platforms could also be used to retransmit data from ground-based sensors in remote areas [7].

　　However, it is unlikely that such additional uses could be incorporated on a communications platform without suffering some loss in the platform's communication ability due to the incorporation of extra equipment and antennas. Sensors also come with considerable additional space and weight requirements, as stabilization systems are often several times heavier than the sensors themselves. Additionally, the operation of a multirole platform may be difficult to coordinate if the other tasks demand an operational profile at odds with its communications tasks. For those reasons, it is most likely that additional tasks such as EW and remote sensing would be conducted from separate dedicated platforms, although the same type of platform could be utilized in each role.

9.4.4 Platform Control

The airborne communications platform must be controlled during its operational period. Options include VHF/UHF direct line-of-sight, UHF satellite, and HF.

Direct line-of-sight

Direct line-of-sight control would require that the ground control station is always in line-of-sight of the airborne platform, thereby limiting the range of the airborne platform. As noted in Section 9.5.1.2, at the highest operating points of 20 km, this range is often limited to 200 km for antennas with a 5° takeoff angle. The requirement for line-of-sight would also invariably imply that the control station and the platform are forward-deployed into the area of operations, increasing system vulnerability and support requirements.

UHF satellite

Satellite command links are perhaps the best option, as the control station can be anywhere within the satellite footprint. One of the advantages of satellite-based command links is that the control station can remain in the strategic environment and the platform does not necessarily have to be forward-deployed. For the Global Hawk UAV, for example, this means that the control station and aircraft base can remain in the strategic environment and reach operating areas that are up to 5,600 km (8 hours transit time) away. This operating range adequately covers those areas required by most countries for tasks in defense of regional interests.

HF.   Communications in this frequency band have limited bandwidth but provide longer ranges than both satellite and line-of-sight command links. It is therefore a useful backup system, provided that there is sufficient space on the platform.

9.4.5 Example Payload

As an example of an airborne platform, the U.S. ACN is based on the Global Hawk UAV and proposes the following services [8]:

- Range extension for 10 to 20 combat-radio channels;
- Range extension for one to three EPLRS channels;
- Range extension for one Link-11 channel;
- Tactical wideband relay for two or four channels for the MSE (trunk subsystem);
- 10 to 20 channels for surrogate tactical satellite;
- Tactical broadcast capabilities for up to 64 Kbps to 1.554 Mbps;
- Internet services for up to 600 users;
- Cellular or PCS telephony for up to 200 calls.

9.5 Some Limitations

Despite its major advantages, the use of an airborne platform does have a number of operational limitations.

- Communications ranges are limited by terrain and weather effects on radio-wave propagation between terrestrial terminals and the airborne platform. These issues affect command links as well as communications circuits.
- The platforms are also vulnerable to physical and electronic attacks, which have the potential to significantly impact the operational use of the assets.
- A number of technical challenges must be overcome if airborne communications platforms are to achieve their potential.

9.5.1 Terrain Effects on Propagation

9.5.1.1 Terrain Screening

The footprint coverage of Figure 9.4 assumes that the Earth's surface is not disturbed by terrain. If ground terminals are located on high ground, communications ranges can be extended beyond 500 km. However, as illustrated in Figure 9.4, the coverage of the airborne platform is limited by the terrain in the near vicinity of the ground terminal. The range of 500 km can therefore vary markedly with the type of terrain.

9.5.1.2 Antenna Elevation Angle

Even if terrain screening is not as marked as that shown in Figure 9.4, at higher frequencies the required antenna elevation angle can limit the range of communication to the airborne platform. In particular, SHF antennas have a radiation pattern that requires clearance above the Earth's surface, which reduces ranges for certain airborne applications such as when it provides a surrogate satellite transponder. Figure 9.5 shows the typical range-extension radius that could be expected for an airborne platform when communicating to ground terminals with antennas that have elevation angles of 0°, 5°, 10°, 20°, and 30°. An elevation angle of 5° should be considered to be the minimum for an antenna at sea level. The problem is not so marked for VHF and UHF whip antennas, which tend to have an omnidirectional radiation pattern.

9.5.2 Effects of Weather on Propagation

While weather will have a small effect on propagation to and from the airborne platform, that effect is either negligible or easily accommodated. At frequencies above 1 GHz, propagation to and from the airborne platform will suffer attenuation by rain and cloud as the radio wave passes through the

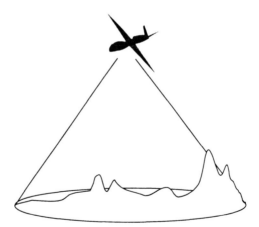

Figure 9.4 Effect of terrain screening on propagation range.

troposphere. However, most communications frequencies for the combat radio and tactical trunk subsystems will be below 1 GHz and will therefore not suffer any additional loss due to weather. Frequencies above 1 GHz (e.g., for surrogate satellite communications) will suffer some loss (on the order of 1–2 dB) and will therefore require slightly larger transmit-powers and higher-gain antennas. Heavy rain may therefore cause some reduction in data rate. However, these difficulties will be much less than those experienced for satellite communications since the airborne platform is 20 to 2,000 times closer than communications satellites.

9.5.3 Tactical Vulnerability

One argument often targeted against airborne communications platforms is their supposed vulnerability against air attack. For example, attrition rates for UAVs are generally accepted to be higher than for manned aircraft. The performance of the air vehicle is optimized in terms of cost, size, and weight, and there is little room for system redundancy within the airframe. The kill probability of the air vehicle, when hit, is very high. However, it appears that

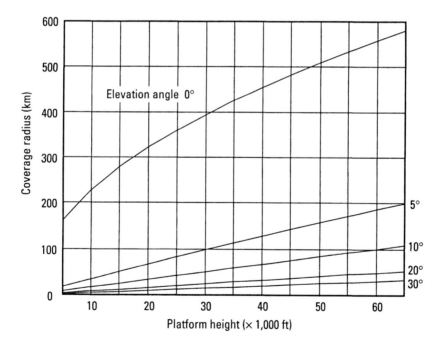

Figure 9.5 Airborne range extension radius as a function of height, for various antenna elevation angles [9].

a UAV is difficult to locate as it presents a small visual profile as well as a small RCS and is therefore difficult to hit either with gunfire or a missile. Aerostats in particular present a difficult target, as two Canadian F16s recently found to their chagrin when trying to shoot down a wayward aerostat that had wandered into controlled airspace. At greater heights such as 20 km, where conventional fighter aircraft struggle to operate, an airborne platform is relatively safe against conventional attack.

The platform is also potentially vulnerable to electronic attack of its communications platform, command links, or navigation systems. Measures can be taken to minimize vulnerability to electronic threats, and in this regard the vulnerability of the airborne platform can be reduced as far as possible for any other electronic battlefield asset. However, by virtue of its height, an airborne platform is more vulnerable than terrestrial assets. Adversary EW assets are invariably terrain-limited and would find the elevated airborne platform an easier target for both electronic support measures (ESM) and electronic attack.

These issues are generally irrelevant when the operational environment is considered. An airborne platform has its greatest utility in extending communications ranges when the supported force is dispersed. If the force is dispersed, then it must be assumed that air superiority has been attained, in which case the airborne platform is not vulnerable. The corollary is that, if air superiority has not been attained, the force will not be dispersed and the requirement for an airborne platform is not so great because communications ranges are much shorter.

9.5.4   Technical Challenges

All the RF systems that are intended to be mounted on an airborne platform create considerable spectrum conflict. For example, the U.S. ACN proposes to have 20 combat radio channels when only four can currently operate simultaneously. In all, the platform must deal with more than 75 collocated transceivers with 17 different waveforms, with the entire capability designed to fit into 100 cubic feet, weigh no more than 450 pounds, and consume only 5 kW of power [8].

Other problematic issues are antenna design for an airborne platform to provide steerable arrays of spot beams; antenna stabilization; provision of adequate power; suitable airship structures; station keeping; the provision of control, switching, and networking services; management of encryption algorithms and keys; and size and weight restrictions, which will only be met with the development of smaller, lower-power electronics.

9.6 Payload Architecture

Figure 9.6 shows a proposed upper-level view of the payload architecture onboard the airborne platform. The payload comprises four main elements:

- *Satellite reach-back communications:* This SHF satellite communications system provides data (and potentially control) communications from the AO to the strategic network.

- *Platform control communications:* The airborne platform can be controlled by either UHF satellite communications or direct line-of-sight radio.

- *High-speed bus, switching, and control:* The airborne platform provides a high-speed data bus to interconnect each supported communications system. Also onboard is a high-speed switch, probably based on ATM technology. This element also contains the communications control module (CCM), which is capable of changing the operating parameters of the onboard communications systems.

- *Supported communications systems:* These elements provide the services to supported communications subsystems, including combat radio and trunk radio rebroadcast, division/brigade theater broadcast, surrogate satellite communications, Link-16 repeater, and cellular base station.

The tactical airborne subsystem must include a CCM that is capable of changing the operating parameters of the onboard communications systems (e.g., switching frequencies, hop sets, and cryptographic variables). These parameters should be able to be preprogrammed before deployment as well as by remote control from the ground during operation.

An onboard CCM would increase the number of gateways to provide truly seamless connectivity to link users of dissimilar systems. The CCM is perhaps the most crucial capability for proper operation of the tactical airborne subsystem. The CCM must perform the following functions:

- Provide interconnection between all onboard communications services and equipment;

- Manage in-flight service priorities, frequency assignment, net initialization, communications security (COMSEC) key assignment, and antenna pointing;

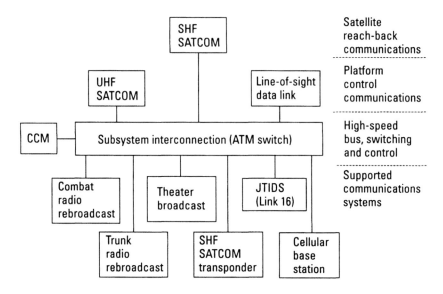

Figure 9.6 Upper-level view of the tactical airborne subsystem's payload architecture.

- Dynamically access and reallocate unused communications channels on the airborne platform;
- Provide gateway connectivity and data format conversions between subsystems.

9.7 Potential Platforms

There are three main potential platforms for the tactical airborne subsystem: piloted aerial vehicles, UAV, and aerostats (balloons). While piloted airborne communications platforms have been employed during World War II and other conflicts such as the Vietnam War, UAV and aerostat platforms have generally been considered somewhat eccentric. However, a number of commercial consortia—such as Angel Technologies, Sky Station, Astrolink, EuroSkyWay, SkyBridge, and Teledesic—are currently developing systems in competition with broadband satellite-based projects [10].

9.7.1 Piloted Aerial Vehicles

The most conventional solution is to mount suitable communications packages in a piloted platform. As an example, the following is a brief description of the Angel Technologies' High Altitude Long Operation (HALO) system, which comprises a fleet of specialized Proteus aircraft that carry broadband transponders communicating through a large underbelly antenna. The piloted aircraft is designed to orbit for up to 16 hours at a time at approximately 16 km, providing a variety of fixed and mobile wireless services, including voice, data, and video. Plans are well advanced, with an August 1999 trial aircraft achieving a 52-Mbps wireless link with the ground [11]. On November 1, 2000, the Proteus aircraft set three world records in peak altitude (62,786 feet), sustained altitude in horizontal flight (61,919 feet), and peak altitude (of 55,878 feet) with a 1,000-kg payload [12].

9.7.1.1 Angel Technologies' HALO System

As an example of a solution based on a piloted vehicle, this section briefly outlines Angel Technologies' HALO System [13]. Angel Technologies Corporation and its partners—Endgate Corporation, Scaled Composites, and Wyman-Gordon Company—propose a piloted HALO aircraft with a fixed-wing airframe and twin turbofan propulsion (see Figure 9.7) that will operate above 51,000 to 60,000 feet over selected cities.

The HALO Proteus aircraft (manufactured by Scaled Composites, a division of Wyman-Gordon Company) was designed to be certified by the Federal Aviation Administration (FAA) for piloted commercial operation; this aircraft can operate from any regional airport within a 500-km radius of the city to which communications are provided. The proof-of-concept aircraft first flew publicly in September 1998. Proteus is built from composite materials and has a gross weight of 6.4 tons, including 2.8 tons of kerosene. Seating is available for two pilots (as well as a spare seat for a third pilot or passenger). Two pilots will be used in the short term to streamline FAA approval; in the long term, one pilot, or even an uninhabited vehicle, may be employed.

The aircraft has been designed for an on-station endurance of 14 hours and operation up to a ceiling of 64,000 feet. However, typically, the platform will fly between 51,000 to 60,000 feet and will remain on station for up to eight hours, dictated mainly by pilot endurance. Each HALO location is serviced by a fleet of three HALO aircraft providing one aircraft on-station at any time. The aircraft orbits at a diameter of 9 to 15 km and uses GPS to maintain its position within 100m.

The aircraft will carry nearly 900 kg of payload, including an antenna array and electronics, in a large streamlined pod underneath. The pod is roll-stabilized so that it swivels to remain parallel to the Earth as the aircraft banks. The antenna array creates hundreds of contiguous virtual cells on the

Figure 9.7 The HALO Proteus aircraft. (Photograph courtesy of Raytheon.)

ground to serve thousands of users. The liquid-cooled payload can be provided with greater than 40 kW of direct-current power.

The ground stations are relatively simple, although a steerable antenna is required because the ground station must track the aircraft that is orbiting at a diameter of 9 to 15 km.

The HALO aircraft will provide the hub of a star-topology broadband telecommunications network that will allow subscribers to access multimedia services, the Internet, and entertainment services as well as to exchange video, high-resolution images, and large data files. Information addressed to non-subscribers or to recipients beyond the region served by the HALO network will be routed through a dedicated HALO gateway connected to the public-switched networks, or via business premise equipment that is owned and operated by service providers connected to the public networks.

The communications payload will operate in two 300-MHz portions of spectrum of the 28-GHz band. The total capacity of one platform can be in the range of 10 to 100 Gbps. Individual consumers will be able to connect at rates ranging from 1 to 5 Mbps and business users will be offered connection speeds ranging from 5 to 12.5 Mbps. A dedicated beam service can also be provided to those subscribers requiring 25 to 155 Mbps. Angel Technologies claim that between 10,000 to 75,000 1.5-Mbps channels can be supported within an area of approximately 100 km in diameter.

9.7.2 UAV

There is considerable worldwide development of UAV-based platforms, predominantly for surveillance operations. UAV-mounted sensors on platforms such as Predator and Phoenix have provided valuable information in theaters such as Bosnia and Kosovo. Attention is now turning, particularly in the United States, to the employment of UAVs in communications and EW modes.

Commercial plans for UAV platforms include the U.S. Helios [14] platform, which is planned to carry 600-pound payloads to 70,000 feet; and the European Heliplat [15] aircraft, which is planned to have a 70-m wingspan. Both aircraft are solar-powered, using fuel cells to provide power at night.

9.7.2.1 U.S. ACN

An example of a UAV-based communications solution is the U.S. ACN concept, which proposes to use the Global Hawk aircraft as a platform [16]. The objective ACN system will provide communications capabilities to support existing joint and Army communications architectures, and to correct communications deficiencies identified during recent deployments and operations.

The ACN combines the capability of a high-altitude endurance UAV with the essential capabilities of a state-of-the-art communications package (node). The ACN's capability to self-deploy anywhere in the world will free up airlift assets that can be used for other missions. The ACN will carry robust communications packages that can be reconfigured rapidly to support changing C2 priorities. High-gain antennas, coupled with the ACN's ability to loiter at very high altitudes (65,000 feet and higher) for extended periods of time, will enable tactical users equipped with lightweight omnidirectional antennas and low-powered radios to establish over-the-horizon communications from mobile platforms. This capability will provide a significant improvement in C2 on the move.

The ACN will be a uniquely capable platform for greatly improving battle command and battle management communications. The ACN's lift capacity will allow it to carry a large, multiband, multimode, and robust communications payload to support a relatively large number of subscribers.

The capability to operate at high altitudes will provide a large communications footprint diameter of 400 to 650 km (depending on ground systems deployed). Loiter endurance times of 24 to 96 hours or more will help ensure user access. A robust antenna and power suite will support the integration of leap-ahead information and communications technology as it becomes available, without major modifications to the airframe, antennas, or power bus. As illustrated in Figure 9.8, the objective ACN will be fully modular, with a common power and signal bus, and a flexible antenna system. This will allow rapid reconfiguration between missions.

As an essential part of the Warfighter Information Network (WIN), the ACN contributes to the rapid connectivity of the entire network. The force projection capabilities of the ACN will greatly enhance the communications capabilities of the WIN architecture. The ACN will provide reachback connectivity from the area of operations to sustaining bases. It will also provide gateways for seamless communications between dissimilar communications systems. The ACN will provide communications redundancy to ensure information dominance, reduce the requirement for terrestrial line-of-sight radio relays, and provide new types of communications services directly to the warfighter.

Although the ACN augments commercial and military satellites, it does not replace them. Satellites and satellite radio systems operate in specific frequency bands and provide unique communications services to support WIN. Satellites, however, will not have the capability to support range extension for every type of military radio and communications system. Range extension for voice and data line-of-sight communications through the ACN will enhance the warfighter's C2 capability.

The ACN system will provide communications capabilities to support existing joint and Army communications architectures, and to correct communications deficiencies identified during recent deployments and operations. The ACN capability should include:

- A modular communications node payload with gateway capability to support and interconnect deployed forces that have both legacy and state-of-the-art communications systems;

- A robust antenna system, versatile power suite, and modular communications package to support rapid reconfiguration on a mission-by-mission basis, as communications priorities change;

- Range extension (retransmission) of SINCGARS, EPLRS, UHF surrogate satellite, MSE, and JTIDS, as well as a limited gateway

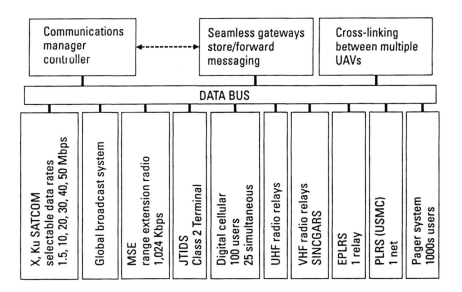

Figure 9.8 Objective capabilities for U.S. ACN [17].

capability between SINCGARS and UHF satellite; range extension implies the following requirements:

- The ACN must be capable of performing retransmission for CNR/SINCGARS nets. An airborne gateway between SINCGARS and UHF single-channel TACSAT equipment will provide on-the-move capability for mobile platforms, without the need for large, directional auto-track TACSAT antennas.

- The JTIDS provides support to near-real-time air-defense engagement operations, and its relay capability will extend connectivity between widely dispersed terrestrial air-defense and joint air-battle elements. The ACN will augment other JTIDS relay, including the airborne warning and control system (AWACS), in providing JTIDS range extension.

- The ACN relay will link EPLRS enclaves that are beyond line-of-sight.

- MSE relay will connect widely dispersed signal nodes on the battlefield.

- A reach-back capability for digital cellular phones, to support early entry and major offensive operations—cellular phone services may be limited because commercial cell sites are large and heavy, and there is at present no incentive for the commercial industry to downsize; reach-back communications will consist of a satellite link with minimum data rate of 1.544 Mbps;

- A communications control element (CCE) capable of switching frequencies, hop sets, and crypto-variables by remote control from the ground; an onboard capability will allow for preprogrammable frequencies, hop sets, and crypto-variables, as well as over-the-air rekeying of hop sets and crypto-variables;

- An onboard communications manager/controller (CMC)—perhaps the most crucial capability for proper operation of the ACN—to increase the number of gateways to provide truly seamless connectivity to link users of dissimilar systems and which must perform the functions described earlier for the CCM (Section 9.6);

- Range extension for Army and joint videoconferencing;

- LPI/LPD communications to store and forward e-mail for SF and conventional elements that operate deep in enemy territory;

- A Global Broadcast System (GBS) relay to provide broadcast communications on the move to support tactical users with omnidirectional antennas and low-powered receivers; global messages such as NBC warnings can also be uplinked to the ACN and rebroadcast to miniature pagers on the ground;

- A flyaway transit case launch and recovery element that can deploy early in a contingency operation and allow in-theater mission controllers to direct the can;

- Cross-linking between multiple ACNs to allow the facility to cover an entire theater of operation;

- Operation in secure and nonsecure modes;

- Interoperability with joint architectures.

As an essential part of WIN, the ACN contributes to the rapid connectivity of the entire network. The force projection capabilities of the ACN will greatly enhance the communications capabilities of the WIN architecture. The ACN will provide reach-back connectivity from the area of operations to sustaining bases. It will also provide gateways for seamless communications between dissimilar communications systems. The ACN will provide communications redundancy to ensure information dominance, reduce the requirement for terrestrial line-of-sight radio relays, and provide new types of communications services directly to the warfighter.

9.7.3 Aerostat Platforms

An alternative to UAV is provided by *aerostats*, which are also called *dirigibles*, or *balloons*. Several commercial consortia propose to use aerostats as the platforms for metropolitan coverage for broadband services. For example, Sky Station International plans to have 250 helium-filled balloons hovering 21 km above the world's biggest cities by 2005. Similarly, the Japanese Ministry of Posts and Telecommunications has advanced plans for an aerostat-based broadband access network, called the Stratospheric Wireless Access Network [18].

Another system is being developed by Turin Polytechnic in conjunction with the Italian Space Agency (ISA) to provide a long-endurance platform that will be powered by solar energy during the day and fuel cells at night [11]. The Korean Ministry of Commerce, Industry, and Energy has plans for a multipurpose stratospheric airship to be used for remote sensing and telecommunications applications [19].

Tethered aerostats. Aerostats may be tethered, that is, tied to the ground. Tethered aerostats have found particular application in elevating sensors and are used in surveillance tasks in such locations as U.S.-Mexico and Iraq-Kuwait borders. In the commercial environment, tethered aerostats are also finding increasing application. For example, Platform Wireless International Corporation plans to use a tethered 1,250-pound airborne payload to provide cellular telephone service to a 140-mile-diameter region in Brazil [20]. However, tethered aersotats tend to have limited ranges due to their low altitudes of around 17,500 feet (constrained by the technology of the tethering cable and weather conditions).

Disposable balloons. Australian Defence Science and Technology Organisation (DSTO) scientists have demonstrated the feasibility of releasing a low-cost ($1,000), disposable meteorological balloon carrying a small transponder to provide short-term (2–3 hours), over-the-horizon communications. This concept was originally developed for the Royal Australian Navy, but it has promise for tactical units to extend VHF/UHF coverage temporarily.

As an example of an aerostat-based solution, Section 9.7.3.1 briefly describes Sky Station's Stratospheric Telecommunications Service [21] concept.

9.7.3.1   Sky Station's Stratospheric Telecommunications Service

The Sky Station's uninhabited aerostat platforms are planned to be guided under remote control on an ascent path to the stratosphere, where GPS is used to locate the platform in the desired geostationary position. The platform then remains in that position for up to 10 years providing high-speed, high-capacity wireless broadband services to an area of approximately 19,000 km^2. Remote sensing and monitoring devices can also be installed on the platform, providing continuous data collection.

The $7.5 billion plan is to offer wireless communication services to users of laptops and handheld terminals at speeds of 64 Kbps to 2 Mbps. The stations will probably accommodate anywhere from 50,000 to 150,000 communications channels, although the theoretical limit for the network is 650,000 64-Kbps channels, according to Sky Station [22].

Ultimately, Sky Station plans to have at least 250 Sky Station platforms, one above every major city in the world. Additional platforms may be located above large population centers, such as Tokyo or London, and additional platforms can be added at any time to increase capacity over specific regions.

Platforms.   The size of Sky Station platforms depends on market demand or the services onboard. The average platform will be approximately 157-m long and 62m in diameter at the widest point. The platform is equipped with sufficient solar and fuel cell capacity to carry a payload of up to 1,000 kg. Although designed for a lifespan of 5 to 10 years, platforms can be recalled for repair if necessary. A new platform will be deployed in advance to replace the existing one so there will be no interruption of service. Catastrophic rupture of the main hull is unlikely due to the use of state-of-the-art envelope materials and weaves. However, in the event of a loss of buoyancy, the automated master control system can be enabled to propel the platform safely to a water landing.

Communications services.   Sky Station's stratospheric platforms will be ideally suited to delivering telecommunications services. Broadband and mobile communications services can be provided at low cost with low latency (less than 0.5 ms compared to about 250 ms for GEO-based services). Additionally, for mobile communications no handover is required. Handover is a significant design issue in terrestrial systems and causes many problems in LEO-based services.

Broadband services. The Stratospheric Communications Service promises to provide cheap, easy access to broadband services in direct competition to the more expensive terrestrial and satellite-based solutions. Subscribers transmit directly with the platform, where onboard switching routes traffic directly to other Sky Station subscribers within the same platform coverage area. Traffic destined for subscribers outside the platform coverage area is routed through ground stations to the public networks or to other platforms serving nearby cities. With data rates of up to 2-Mbps uplink and 120-Mbps downlink, subscribers will be provided with high-speed Internet access, as well as other broadband services such as television distribution, videoconferencing, and on-line remote monitoring and other security applications. Spectrum in the 47-GHz band (47.2–47.5 GHz stratosphere-to-Earth and 47.9–48.2 GHz Earth-to-stratosphere) has already been designated globally by the ITU and by the U.S. Federal Communications Commission (FCC) for use by stratospheric platforms.

Mobile communications. The Sky Station system is also the ideal means for low-cost rapid deployment of mobile services. Sky Station is participating in the development and delivery of a third-generation cellular service.

9.8 Costs

Current proposals for commercial stratospheric systems anticipate costs of between $50 million and $1 billion as opposed to approximately $9 billion for an LEO satellite constellation. In the absence of detailed user requirements, accurate costs are currently difficult to estimate and compare for each potential platform. However, the following indicative costs allow some basis for comparison.

The U.S. ACN is currently under design by three U.S. teams headed by Raytheon, Sanders, and TRW. DARPA believes that a full flyaway capability can be achieved for less than $5 million per aircraft unit. Taking into account full operating costs, Australian estimates for Global Hawk life-cycle costs (one platform and ground equipment) are approximately $1 billion in Australian dollars over 10 years (acquisition cost and 7 years of operating costs) [23]. For a smaller platform, estimates for 13 Predator-based platforms are $18 million over fiscal years 1997 to 2002 [24].

Although difficult to quantify precisely, costs for aerostat-based solutions are expected to be at least one-half that of UAV solutions, depending on the desired mission profiles. For example, the total investment for the

Turin Polytechnic/ISA platform is expected to amount to only about $3 million and the operating cost might be as low as $345 per hour. This investment is at least an order of magnitude lower than land-based terrestrial broadband distribution systems and two orders of magnitude less than that required for satellite-based platforms [11].

9.9 Conclusions

The tactical airborne communications system provides a significant improvement in communications ranges by extending the combat radio, tactical trunk, and tactical data distribution subsystems. In addition, it allows for command and control on the move across an AO with a radius of between 200 and 500 km. The deployment of an airborne platform offers additional opportunities in the provision of other communications services such as theater broadcast and PCS.

Despite the need to resolve a number of technical challenges, there are a number of potential subspace platforms that will be able to meet the requirements of a tactical airborne subsystem in support of land operations. The precise platform type and payload will depend of the nature of the operational requirement, but it is considered that the larger systems have the most potential due their larger capacities and higher operating altitudes, leading to longer ranges for communication range extension.

Endnotes

[1] Meyer, C., *Division-Level Communications 1962-1973*, Department of the Army: Washington, D.C., 1982.

[2] Bergen, J., *Military Communications A Test for Technology*, Center for Military History, United States Army: Washington, D.C., 1986, pp. 142, 156, and 285.

[3] Bergen, J., *Military Communications A Test for Technology*, Center for Military History, United States Army: Washington, D.C., 1986, p. 164.

[4] Bem, D., T. Wieckowski, and R. Zielinski, "Broadband Satellite Systems," *IEEE Communications Surveys*, Vol. 3, No. 1, 2000, available on-line at http://www.comsoc.org/pubs/surveys/1q00issue/zielinski.html.

[5] This table is based on the similar UAV requirements of *Warfighter Information Network (WIN) Master Plan*, Commander, U.S. Army Signal Center and Fort Gordon: Fort Gordon, GA, Version 3, June 3, 1997.

[6]   McAllister, M., and S. Zabradac, "High-Altitude-Endurance Unmanned Aerial Vehicles Pick Up Communications Node," *Army Communicator*, Spring 1996, pp. 21–23.

[7]   Airborne relays were used for this purpose by U.S. forces in Vietnam; see, for example: Bergen, J., *Military Communications A Test for Technology*, Center for Military History, United States Army: Washington, D.C., 1986, pp. 392–393.

[8]   Ackerman, R., "Defense Department Researchers Aim for Sky-Based Switchboards," *Signal*, April 1999, pp. 65–67.

[9]   Mahoney, T., Annex A to M. J. Ryan and M. R. Frater, *Battlespace Communications System (Land) Architecture Study*, ADF Contract CAPO 3726441, Oct. 1999.

[10] In addition to the references cited in the following sections, further information can be found in:

Djuknic, G., J. Friedenfelds, and Y. Okunev, "Establishing Wireless Communications Services Via High-Altitude Aeronautical Platforms: A Concept Whose Time Has Come?" *IEEE Communications Magazine*, Sept. 1997, pp. 128–135.

Kopp, C., "Angels, HALOs and Atmospheric Networks," *Australian Communications*, Sept. 1999, pp. 23–30.

[11] Edwards, T., "More Than Just Hot Air," *Communications International*, Sept. 1999.

[12] See Scaled Composites LLC press release at http://www.scaled.com/news/pr110100.htm.

[13] Unless otherwise noted, the information in this annex is based on details located at the Angel Technologies' Web site: http://www.broadband.com/ and Collela, N., N. Martin, and I. Akyildiz, "The HALO Network," *IEEE Communications Magazine*, June 2000, pp. 142–148.

[14] More information can be obtained in Zorpette, G., "Winging Wildly Upward," *IEEE Spectrum*, Sept. 2001, pp. 22–23; and at the Web site http://www.dfrc.nasa.gov/ Projects/Erast/helios.html.

[15] Further information is available at http://www.elec.york.ac.au/comms/haps.html.

[16] This annex is largely an extract of the ACN description in *Warfighter Information Network (WIN) Master Plan*, Commander, U.S. Army Signal Center and Fort Gordon, Version 3, June 3, 1997.

[17] *Warfighter Information Network (WIN) Master Plan*, Commander, U.S. Army Signal Center and Fort Gordon, Version 3, June 3, 1997.

[18] More information can be found on the Web site http://www2.crl.go.jp/mt/b181/research/spf/index-e.html.

[19] Announced at the Seoul Air Show 2001—see http://seoulairshow.com/Eng/News/Article/article_14.htm.

[20] Brewin, R., "Giant Aerostats Developed For Rural Cell Phone Service," March 5, 2001, available on-line at http://www.computerworld.com/cwi/stories/0,1199,NAV47-68_STO58321,00.html.

[21] Unless otherwise noted, the information in this section is based on details located at the Sky Station Web site: http://www.skystation.com.

[22] http://www.teledotcom.com/0197/features/tdc0197satelliteside1.html, accessed Jan. 1997.

[23] Hale, G., "ADF Employment of the Global Hawk Uninhabited Aerial Vehicle (UAV)," Air Power Studies Center, Paper 76, July 1999.

[24] Robinson, C., "High-Capacity Aerial Vehicles Aid Wireless Communications," *Signal*, April 1997, pp. 16–20.

10

Tactical Network Interfaces

10.1 Introduction

Previous chapters have described the tactical trunk subsystem, the tactical data distribution subsystem, and the combat radio subsystem. This chapter describes the internal interfaces between these subsystems within the tactical communications system and the external interfaces between the tactical communications system and supported systems, overlaid communications systems, and strategic communications system. These interfaces are the key to the future of the tactical communications system, because they are the glue that will allow a network to be built from the previously separate subsystems.

It has been shown in previous chapters that because there are differing requirements for mobility and capacity, it is not feasible with current or foreseeable technology to provide the tactical communications system as a physically homogeneous network. It is essential, nonetheless, to provide a system that forms a single logical network to facilitate the movement of data throughout the battlespace. This is in line with current trends in commercial networking technology. The interfaces between the different parts of the tactical communications system and between it and other systems play an important part in this integration.

Within the tactical communications system, the following interfaces are required:

- Combat radio subsystem—tactical trunk subsystem;

- Combat radio subsystem—tactical data distribution subsystem;
- Tactical data distribution subsystem—tactical trunk subsystem.

Because the airborne subsystem carries communications equipment from the other subsystems, it does not require an explicit interface. This is discussed further in Section 10.5.4.

The tactical communications system should provide external interfaces to:

- Supported systems, such as command elements (including user terminals, both individually and aggregated), sensors, and information systems;
- The strategic communications system;
- Overlaid communications systems, such as the PSTN, theater broadcast systems, and PCSs.

The local subsystem will provide the interface between supported systems and the tactical communications system. It will also provide limited interfaces between the various subsystems. Because one of the primary purposes of the local subsystem is to provide interfaces, we do not consider explicitly the interfaces between it and other systems.

This chapter begins with a general consideration of the basic requirements of interfaces, followed by an overview of the local subsystem. The more detailed structure of the interfaces nominated above is then considered in turn. The chapter concludes with an overview of the remaining issues that require further study.

10.2   Basic Requirements

Like other parts of the tactical communications system, the interfaces, both internal and external, have a number of basic requirements.

Range/capacity/mobility trade-off.   The interface should not restrict capacity or mobility beyond the restriction of the less capable of the two systems being interfaced. As a result, there may be physical differences in the systems providing the interface in, for example, a brigade headquarters and a company headquarters. The interface may also enhance the communications services by providing buffering between the systems it is connecting. For a messaging service, this buffering might enable a store-and-forward capability.

QoS. An interface should support the full range of communications services provided by connecting entities, and should not itself limit these services. This means that the interface should support the Qos requirements of these services. In areas where a mixture of real-time (i.e., delay-constrained) and non-real-time (i.e., loss-constrained) services are carried, measures are required to support the requirements of both. In addition to the traditional sources of real-time services, some types of cryptographic systems (particularly bulk-encryption systems that produce a continuous output bitstream) may have strict requirements on both loss and delay.

Flexibility. A user should be able to connect the same terminal to the tactical communications system at any interface regardless of location with minimal variation in the procedures used. The only major difference that is visible to the user should be the variation in capacities across the network. There should also be minimal differences between interfaces provided for concentrated forces to those provided for dispersed forces.

Seamless connectivity. An interface should not limit the connectivity provided by the tactical communications system. Interfaces should support the services offered by the tactical communications system, including the naming and addressing scheme adopted by the network. This implies that the interface is capable of performing as a gateway (i.e., an OSI Layer 7 interface) where required.

Simplicity. Interfaces will often be operated by personnel who are not specialist communicators. It is therefore necessary that use of an interface does not require special skills and training.

Precedence and preemption. Support for precedence and preemption at interfaces is especially important to resolve contention for the use of limited resources.

Security. Interfaces must support the security architecture of the tactical communications system. Interfaces may therefore provide encryption and decryption services. They may also provide support for services such as over-the-air rekeying of cryptographic equipment.

Power source. An interface should be capable of being operated from the same type of power source as other elements of the tactical communications system operating in the same location.

Operation in all geographic and climatic conditions.   An interface should be capable of being operated under all conditions of immersion in water, temperature, and extremes of vibration, shock, pressure, and humidity as the systems that are being interfaced.

10.3   User Terminals

Communications services are provided to users through a user terminal, which is interfaced to the tactical communications system. A user of the communications services offered by the tactical communications system may be a staff officer, who may have a fixed location in a headquarters or logistics area, or who may be mobile; a commander (squad, platoon, company, battalion, brigade, or division); or an equipment operator or network manager.

The user's terminal will take different forms, depending on the user's role and location and the communications services required, as illustrated in Table 10.1. It is desirable, however, that each user accesses all network services through a single terminal. In the future, many users will require terminals incorporating both voice and data services. User terminals may take the form of a telephone; a facsimile; a radio handset capable only of voice operation; a data terminal, including palm-top computers, notebook computers, and more powerful desktop and vehicle-mounted computer systems, as well as peripherals such as printers; or a multimedia terminal, incorporating voice, facsimile, data, and possibly videoconferencing.

A user terminal should be able to operate to any part of the tactical communications system with minimal variation in procedures. A user terminal will be capable of interfacing directly to one or more of the tactical trunk subsystems, the tactical data distribution subsystems, the combat radio subsystems, and the local subsystems (described in Section 10.4). It is preferable that user terminals maximize the flexibility of their interfacing capability. A portable terminal should preferably be able to interface to all of the subsystems listed above. In areas where a terminal is permanently mounted in a vehicle and connected to a local subsystem, the capability of interfacing directly to other subsystems may not be required. The deployment, operation and management of a user terminal should be an individual responsibility. Its use should not, therefore, require specialist training.

10.4 The Local Subsystem

Aggregation of traffic from user terminals may occur in a local area such as a headquarters. The system performing this function will be referred to as the

Table 10.1
Possible Types of User Terminals and Their Associated Users

Terminal Type	Services	Typical User
Radio handset	Voice only	Section, platoon, or company commander; staff officers up to brigade; for those users working from vehicles, likely to be integrated into multimedia terminal in the future
Telephone	Voice handset, possibly with data adaptor for data terminal	Staff officers and commanders at battalion and above; is likely to be integrated into multimedia terminal in the future
Facsimile	Text and graphics	Staff officers and commanders at battalion and above; is likely to be integrated into multimedia terminal in the future
Data terminal, varying from personal digital assistant (PDA) to desk-top computer	Data, including text, graphics, and imagery	Staff officers and commanders at all levels; for users at battalion and above, the data terminal is likely to be integrated into multimedia terminal in the future
Multimedia terminal	A wide range of voice and data, supplying all services provided by other types of terminals, but at the cost of higher cost and complexity	Staff officer or commander supported by vehicles, most likely at battalion and higher

local subsystem. The local subsystem may perform a number of functions, including a communications network for a local area, capable of carrying real-time services such as voice telephony and non-real-time services such as data; interfaces to one or more of tactical trunk subsystem, combat radio subsystem, strategic communications system, or overlaid communications systems for users in a local area, depending on available equipment and level; an internal switching capability so that traffic whose source and destination terminals are both connected to the same local subsystem can be passed directly without traveling over the tactical communications system; and a routing capability where it is connected to more than one external system, which would provide an alternate to the primary routing provided by the tactical communications system itself [1].

Through these functions, the local subsystem allows a single user terminal to access the full range of communications resources available in the local area without requiring a separate connection to each one. It may be desirable in some circumstances that the local subsystem provides wireless connectivity for user terminals. Depending on the range over which this interface is required to operate, it might be achieved using an infrared WLAN or an RF WLAN, possibly based on commercial standards such as IEEE 802.11 [2] or Bluetooth [3]. Infrared and Bluetooth-based technologies are likely to be suitable for ranges of a few meters, while commercial WLAN standards such as IEEE 802.11 may be usable over distances of tens of meters.

While the local subsystem sometimes provides switching, this capability should be used primarily for local switching, but is available as an alternative to the switching offered by the tactical trunk subsystem. If the local subsystem is used to provide the primary switching for the tactical communications system, the mobility of users will be unduly restricted.

The deployment, operation, and management of a local subsystem are the responsibility of the unit that it supports. Its use should not, therefore, require specialist training, but should be able to be incorporated into the training cycle of individual, subunit, and unit training.

The simplest local subsystem, illustrated in Figure 10.1, may be a switch that enables a single voice handset or palm-top computer to access two radios—VHF and HF.

Within a small headquarters, the local subsystem may enable a staff officer or commander to use a single handset for accessing voice radios and

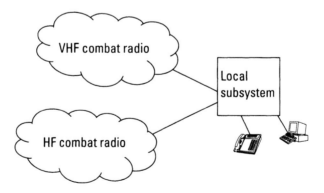

Figure 10.1 Simplest local subsystem.

for making local telephone calls or telephone calls via the tactical trunk subsystem. The local subsystem may also provide an alternate for routing between nets in the combat radio subsystem, and the interface between the combat radio subsystem and the tactical data distribution subsystem.

In a larger headquarters, as illustrated in Figure 10.2, the local subsystem may support a range of multimedia terminals, allowing users to access the full facilities of all subsystems of the tactical communications system. This local subsystem may provide an access node for the tactical trunk subsystem, and an alternate for routing between nets in the combat radio subsystem, routing and switching in the tactical trunk subsystem, the interface between the combat radio subsystem and the tactical data distribution subsystem, the interface between the tactical data distribution subsystem and the tactical trunk subsystem, and the interface between the tactical trunk subsystem and overlaid communications systems.

The various types of local subsystems are summarized in Table 10.2. The basic requirements of interfaces in the tactical communications system are applied to the local subsystem as follows.

Range/capacity/mobility trade-off. A local subsystem should have sufficient capacity to support the user terminals and communications equipment to which it is connected without restricting the capacity of either. A local subsystem should have the same mobility as the user terminals and those subsystems of the tactical communications system to which it is directly connected.

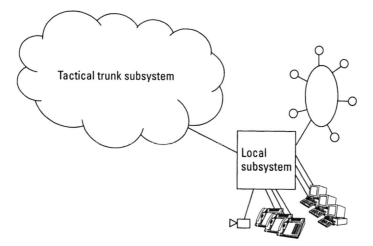

Figure 10.2 Local subsystem in a larger headquarters.

Table 10.2

Types of Local Subsystems

Location	Description
Single vehicle or user	Harness to allow a single terminal to be switched between two or more communications devices (e.g., VHF and HF radio)
Low-level headquarters	Allows use of a single handset for accessing voice radios and making telephone calls; may also provide an alternate for routing between nets in the combat radio subsystem, and the interface between the combat radio subsystem and the tactical data distribution subsystem
Larger headquarters (battalion or brigade and above)	Supports a range of multimedia terminals, allowing users to access the full facilities of network; provides access to subsystems and alternate to various interfaces between subsystems and between the tactical communications system and external systems

QoS. Support for real-time and non-real-time services is required in areas where these are supported either by attached terminals or communications equipment.

Flexibility. While equipment providing the local subsystem may vary due to differing requirements of mobility, capacity, and power source, commonality of interfaces to user terminals and communications equipment should be maximized.

Seamless connectivity. A local subsystem should support the full connectivity supported by subsystems of the tactical communications system to which it is connected. A local subsystem connected to the tactical trunk subsystem, for example, should therefore support voice telephony, while a local subsystem that connects only to the tactical data distribution subsystem may not.

Simplicity. There should be as much commonality as possible between the operations of different types of local subsystems. It is desirable that the interface between the local subsystem and the tactical trunk subsystem be as similar as possible to internal interfaces in the tactical trunk subsystem. Such similarity is desirable in the interfaces to the combat radio subsystem and tactical data distribution subsystem. Simplicity requires also development of a common vehicle harness.

Precedence and preemption. Support for precedence and preemption in the local subsystem enables limited communications resources to be shared between users. This is important both within the local subsystem and at interfaces between the local subsystem and the tactical communications system.

Security. The operation of a local subsystem should not conflict with security requirements. This may require the provision of encryption within a local subsystem where cabling runs over significant distances.

Minimal mutual interference. The local subsystem should not generate any electromagnetic interference (EMI) that inhibits the operation of another interface or any other part of the tactical communications system. This should hold regardless of whether all ports of the local subsystem are in use, or some are unconnected.

Power source. The power source of a local subsystem should be the same as either the attached user terminals or communications equipment.

Operation in all geographic and climatic conditions. A local subsystem should be ruggedized to the same standards as the attached user terminals or communications equipment.

10.5   Internal Interfaces

10.5.1   Combat Radio Subsystem: Tactical Trunk Subsystem Interface

Connectivity between the combat radio subsystem and the tactical trunk subsystem must support the passage of both voice and data.

The interface should facilitate the carriage of voice traffic from the combat radio subsystem into the trunk system. The full facilities of the tactical trunk subsystem should be available to a user connecting from the combat radio subsystem. This would include, for example, the ability to place and receive telephone calls. Similarly, the full facilities of the combat radio subsystem should be available to a user connecting from the tactical trunk subsystem. This would include being able to join a net.

Voice interface between the combat radio subsystem and the tactical trunk subsystem may be limited to only those users of the combat radio subsystem who possess a station on a net that has one of its stations connected directly to the tactical trunk subsystem. This limitation arises because the combat radio subsystem does not normally allow for routing between nets. Access to some voice facilities of the tactical trunk subsystem from the combat radio subsystem may not be available without operator assistance.

For non-real-time data, the interface should permit the passage of messages and other data in both directions. Unless the protocols used in the tactical trunk subsystem and the combat radio subsystem are identical, the interface will need to provide the full functionality of a gateway.

Gateway functions associated with this interface should be part of the trunk communications subsystem, since its mobility requirement is less than that of the combat radio subsystem. The interface is therefore managed as part of the tactical trunk subsystem.

As illustrated in Figure 10.3, in addition to providing a gateway to the combat radio subsystem, the tactical trunk subsystem may also provide a direct bridging capability between nets on the combat radio subsystem. This differs from the gateway function in that it does not require messages and other data to be converted to protocols known to the tactical trunk subsystem. In this situation, the tactical trunk subsystem provides a *virtual private network* (VPN) functionality for the combat radio subsystem.

Every node in the tactical trunk subsystem should be capable of providing an interface to the combat radio subsystem.

10.5.2 Combat Radio Subsystem: Tactical Data Distribution Subsystem Interface

Connectivity between the combat radio subsystem and the tactical data distribution subsystem must support the passage of data. A real-time voice interface is only required if the tactical data distribution subsystem carries this type of traffic. It is expected that this would not normally be the case, as

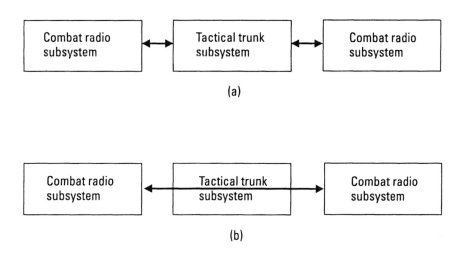

Figure 10.3 (a) Gateway interface and (b) VPN interface.

the primary role of the tactical data distribution subsystem is to support real-time situational awareness.

For non-real-time data, the interface should permit the passage of messages and other data in both directions. Unless the protocols used in the tactical data distribution subsystem and combat radio subsystem are identical, the interface will need to provide the full functionality of a gateway.

Gateway functions associated with this interface should be part of the tactical data distribution subsystem, since it is most likely that at least some elements of the tactical data distribution subsystem are vehicle-mounted, whereas the entire combat radio subsystem may be man-portable. The interface may be managed either as part of the tactical data distribution subsystem or as part of the tactical trunk subsystem.

In addition to providing a gateway to the combat radio subsystem, the tactical data distribution subsystem may also provide a direct bridging capability between nets on the combat radio subsystem. This differs from the gateway function in that it does not require messages and other data to be converted to protocols known to the tactical data distribution subsystem. In this situation, the tactical data distribution subsystem provides a VPN functionality for the combat radio subsystem.

Every node in the tactical data distribution subsystem should be capable of providing an interface to the combat radio subsystem. At small nodes, this interface may be provided by the node's local subsystem.

10.5.3 Tactical Data Distribution Subsystem: Tactical Trunk Subsystem Interface

Connectivity between the tactical data distribution subsystem and the tactical trunk subsystem must support the passage of data. No requirement exists for a voice interface between the tactical data distribution subsystem and the tactical trunk subsystem. This is because the tactical data distribution subsystem does not carry real-time conversational services such as voice.

For non-real-time data, the interface should permit the passage of messages and other data in both directions. Unless the protocols used in the tactical trunk subsystem and tactical data distribution subsystem are identical, the interface will need to provide the full functionality of a gateway.

Gateway functions associated with this interface should be part of the trunk communications subsystem, since its mobility requirement is less than that of the tactical data distribution subsystem. The interface is therefore managed as part of the tactical trunk subsystem.

In addition to providing a gateway to the tactical data distribution subsystem, the tactical trunk subsystem may also provide a direct bridging capability between nets on the tactical data distribution subsystem. This differs from the gateway function in that it does not require messages and other data to be converted to protocols known to the tactical trunk subsystem. In this situation, the tactical trunk subsystem provides a VPN functionality for the tactical data distribution subsystem.

Every node in the tactical trunk subsystem should be capable of providing an interface to the tactical data distribution subsystem.

10.5.4 Tactical Airborne Subsystem Interface

The tactical airborne subsystem is used to extend the range of the tactical trunk subsystem, the tactical data distribution subsystem, and the combat radio subsystem.

This interface creates no new interface requirements. If it extends the combat radio subsystem, it carries equipment from the combat radio subsystem. If it provides an interface between combat radio and tactical trunk subsystems, it will carry equipment from both subsystems, including the same interface as would be used if this interface were provided on the ground.

Two possibilities are foreseen for which a new interface may be required in the future. These are situations in which the tactical airborne subsystem includes a surrogate satellite transponder, in which case different ground-based equipment may be required from that used for conventional ground-ground or ground-air links; and provides direct air-air links between rebroadcast stations.

10.6 External Interfaces

There exist a large number of external systems to which the tactical communications system must interface, including supported systems, the strategic communications system, and overlaid communications systems. This section describes interfaces to these external systems.

10.6.1 Interface to Supported Systems

The tactical communications system is required to support a number of different types of supported systems. These are defined in Chapter 5 as:

- *Command elements, maneuver elements,* and *logistics elements;*
- *Sensor systems,* which would often provide their own communications between sensor and interpretation facility and would be interfaced to the tactical communications system at this facility;
- *Weapons platforms,* which should be able to connect in to any point in the tactical communications system and subsequently be able to access any sensor, the supported command element, and their own command post;
- *Information systems,* which include tactical-level systems, joint and combined systems, and a wide range of administrative systems;
- *Information services,* including security, messaging, video teleconferencing, information management, data replication and warehousing, distributed computing, and search engines;
- *Network management,* which must be an integrated system capable of managing the entire tactical communications system.

As much as possible, interfaces for these supported systems should be provided across the tactical communications system.

This interface may be part of the combat radio subsystem, the tactical data distribution subsystem, the tactical trunk subsystem or the local subsystem. It is managed by as part of the subsystem of the tactical communications system to which the supported system is directly connected.

10.6.2 Interface to Strategic Communications System

It is necessary to connect the tactical communications system to operational- and strategic-level communications systems. This permits the command of deployed forces to be exercised by operational and strategic headquarters. It also allows deployed forces to access communications and information services such as intelligence databases and logistics support systems via the strategic communications system.

The interface to the strategic communications system should support both real-time services such as voice and non-real-time data services. This interface may be provided via a number of types of bearer, including HF radio, satellite, or the public telephone network.

Unlike other types of interface described in this chapter, this interface is usually provided across a radio path.

Every node in the tactical trunk subsystem should be capable of providing an interface to the strategic communications system, given a suitable bearer. This bearer could be the public telephone network, a satellite link, or an HF channel. Nodes in the combat radio subsystem may be capable of providing an interface to the strategic communications system (e.g., via HF radio), subject to the requirement that this interface is provided by specialist units.

Depending on management arrangements, this interface may be achieved by having the boundary between the tactical communications system and the strategic communications system may be see to be "in the air," by placing equipment conforming to tactical communications system standards at strategic sites, or by having equipment conforming to strategic communications system standards at tactical sites.

The interface between the strategic communications system and the tactical communications system should normally be deployed, operated, and managed by specialist signals units. The interface to the strategic communications system should be managed as part of the strategic communications system. Equipment that is collocated with the node of the tactical communications system, however, should be operated by the unit collocated with that node. The interface should facilitate support provided to the tactical communications system by the strategic communications system.

The U.S. *global information grid* (GIG) extends the concept of interfaces to operational and strategic networks, aiming to provide seamless integration throughout a reliable, assured, cost-effective, global network. An important means for providing this level of connectivity will be the incorporation of multiple layers of airborne rebroadcast using aircraft, UAVs, and satellites [4]. The GIG allows operational and strategic networks to act as overlaid communications systems.

The basic requirements of interfaces in the tactical communications system are applied as follows.

Range/capacity/mobility trade-off. The capacity of this link will be determined largely by the capacity of the bearer across which it operates. Buffering should be provided in both the strategic communications system and the tactical communications system to maintain seamless connectivity in the presence of congestion in the interface or the tactical communications system or temporary loss of connectivity in the tactical communications system.

QoS. This interface should support both real-time services such as voice and non-real-time data services.

Flexibility. It is desirable that the differences other than capacity in this interface due to the use of different bearers be minimized.

Seamless connectivity. This interface is required to support the provision of communications and information services across both the strategic communications system and the tactical communications system. The interface should permit services offered by the strategic communications system to be accessed from all user terminals attached to the tactical communications system, subject to sufficient capacity being provided by the bearer and the tactical communications system providing the required connectivity between the user terminal and the interface to the strategic communications system.

Simplicity. While part of this interface resides with a node of the tactical communications system, it should be managed as part of the strategic communications system. A higher level of complexity is therefore permissible than for interfaces that must be managed by users.

Security. The interface must support the security architectures of the tactical communications system and strategic communications system and maximize the ability of services to be provided across both systems.

Precedence and preemption. Precedence and preemption should be supported across the interface.

Minimal mutual interference. This interface should not generate any electro-magnetic interference that inhibits the operation of another interface or any other part of the tactical communications system.

Power source. The interface to the strategic communications system should use the same power source as the node of the tactical communications system to which it is directly connected.

Operation in all geographic and climatic conditions. This interface should be ruggedized to the same standards as the equipment in the node of the tactical communications system to which it is directly attached.

10.6.3 Interfaces to Overlaid Communications Systems

The tactical communications system provides the minimum essential communications for the conduct of land operations. The tactical communications system should be able to make use of these to expand its capacity in areas where other communications systems are available.

10.6.3.1 PSTN Interface

The PSTN provides a circuit-switched network that supports a range of services, including voice, video, and data. The services available depend on the type of connection used. Commonly available connections are a single channel analog interface, that can carry either one voice call or data at rates up to approximately 30 Kbps in ideal operating conditions; an ISDN basic rate interface, which provides two 64-Kbps channels, each of which can carry a single voice call, and a 16-Kbps signaling channel; and an ISDN primary rate interface, providing a data rate of 2 Mbps or 30 multiplexed voice circuits. The PSTN also provides a range of interfaces that support global roaming, including interfaces to PCS and satellite communications systems.

Not all types of interface are available in all locations. ISDN interfaces, in particular, are likely to be available only in major population centers.

The PSTN may be used as an overlaid communications system in order to supplement the capacity of the tactical communications system, connect the tactical communications system to the strategic communications system, and provide connectivity between user terminals in the tactical communications system and terminals attached to the PSTN.

Advantages offered by the use of the PSTN as an overlaid communications system include provision of interconnection between tactical communications system users and locations not served by the tactical communications system, such as civil authorities and civil logistics suppliers;

additional capacity to the tactical communications system; and an alternate channel.

Disadvantages include the fact that capacities available away from major population centers may be small, and the physical vulnerability of the PSTN, especially its cables, which are difficult to protect.

Every node of the tactical trunk subsystem should be capable of providing an interface to the PSTN. This interface is managed as part of the tactical trunk subsystem.

10.6.3.2 Public Data Network Interface

Many countries have public data networks, including the Internet and commercial packet-switching networks based on the ITU-T X.25 standard. There are also international connections between these national networks. Access to these networks can be obtained by dedicated connections or by dial-up connections using a modem.

A public data network may be used as an overlaid communications system to provide additional capacity to the tactical communications system, or act as an alternate channel.

The advantages offered by the use of a public data network as an overlaid communications system include the provision of additional capacity to the tactical communications system and the provision of an alternate channel. Potential disadvantages include the fact that datagram headers cannot be encrypted because they are used in routing decisions made by the network. This prevents bulk encryption at the interface to the tactical communications system or user terminal. Signaling traffic between the network and tactical communications system must also be transmitted without encryption. Consequently, security against traffic analysis is typically difficult to provide, although some security can be provided by the use of VPN technology.

10.6.3.3 Satellite Communications Systems Interface

Communications satellites provide long-distance communications between ground-based stations. In systems currently in service, the satellite may provide signal regeneration, but it does not provide other services such as switching. Military satellites (or military transponders on commercial satellites) may provide EP. The capacity of a satellite transponder depends on the transmitter output power and the gain of the Earth station antenna. The performance of the least-capable terminal tends to limit the total performance that can be obtained in areas where multiple Earth stations share a transponder.

Satellite communications may be used as an overlaid communications system to provide connectivity between geographically dispersed elements of the tactical communications system, access to the strategic communications system from the tactical communications system, access to the PSTN from the tactical communications system, or access to the tactical communications system from an isolated user terminal or local subsystem.

Advantages offered by the use of satellite communications as an overlaid communications system include interconnection between tactical communications system users and locations not served by the tactical communications system, such as civil authorities and civil logistics suppliers; provision of additional capacity to the tactical communications system; and provision of an alternate channel. Disadvantages include the lack of EP in commercial satellite systems, which leads to vulnerability to adversary EA and ES.

10.6.3.4 PCS Interface

PCS may be used as an overlaid communications system. The nature of the interfaces involved is shown in Table 10.3. PCS may be used as an alternative to the laying of line in a headquarters or logistics installation, as in interface to the PSTN or a means of access to the tactical communications system for users who are temporarily outside its coverage. This latter use is likely to be restricted to low-intensity operations, in which civilian infrastructure is less likely to have been damaged.

Advantages offered by the use of PCS as an overlaid communications system include provision of access to communications resources from locations not directly supported by the tactical communications system, and provision of an alternate channel. Disadvantages include the lack of EP in commercial PCS, which leads to vulnerability to adversary EA and ES, especially in base stations and satellites; the presence of a base station required for operation with most systems; and the necessity that mobile stations must be within range of a base station, which is limited to approximately 35 km for TDMA systems based on GSM [5].

In the first case, the PCS is operated, managed, and maintained by an external organization. This is the only case that will typically be applicable to satellite PCS. In the second case, the PCS is operated, managed, and maintained by the deployed force.

10.6.3.5 Theater Broadcast Interface

A TBS, known as the Global Broadcast System in the United States, uses the concept of high-bandwidth broadcasting for providing a high-capacity broadcast service. Military TBS is typically required to be delivered over a long distance. Likely options for the communications channel include

Table 10.3
Interfaces to PCS

Use of PCS	Nature of Interface
Alternative to line within headquarters or logistics installation	Base station interfaces to tactical communications system; PCS handsets (possibly ruggedized) are used for access.
Interface into PSTN	Tactical communications system includes an interface that emulates a PCS handset to PCS base station. User terminal is the same as used with integral communications systems, and is connected to PCS via the tactical communications system.
Access for remote users	Users who are outside the coverage of integral elements of the tactical communications system may use PCS handsets to access the tactical communications system via the PSTN.

satellite, possibly using a high-power steerable spot beam; VHF, UHF, or SHF radio with airborne repeater; or HF sky wave.

A different broadcast may need to be provided by each level of command. One picture may be provided to brigades, while each brigade may filter and refine this picture before passing it on to units under command. This doctrine has yet to be finalized.

A TBS may be used as an overlaid communications system to provide many services that would otherwise be carried across the tactical communications system, including database replication and weather reports. The advantages offered by the use of a military TBS as an overlaid communications system include such diverse elements as high capacity and efficient provision of multicast services. Disadvantages include the performance being limited by the user with the lowest performance antenna and potential difficulty of integration with protocols of the tactical communications system that may require duplex channels.

Some nodes of the tactical trunk subsystem should be capable of providing an interface to TBS. Access to TBS data is then available to users via the local subsystem. This interface is managed as part of the tactical trunk subsystem. It is conceivable that the tactical communications system may be used to pass data received from TBS to elements that cannot receive it directly. The tactical communications system is unlikely to have sufficient capacity to carry large amounts of TBS data, however, and it is therefore more likely in practice that this direct bridging would not occur.

10.7 Conclusion

The provision of suitably designed interfaces in the tactical communications system is one of the major requirements to enable a fully functional battlefield network. In designing a particular system, however, many different requirements must be taken into account, some of which are summarized in Table 10.4. These requirements can be used to form the basis of studies to implement these interfaces, including technology options, costs, and benefits.

Table 10.4
Issues Requiring Study in the Design of Interfaces in the Tactical Communications System

Location	Issues
Local subsystem	Provision and dimensioning of the local subsystem for headquarters at all levels and logistics installations;
	Vehicle harnesses for use in all local subsystems;
	A detailed work breakdown structure and cost analysis for the provision of the subsystem;
	Manning, ILS, maintenance, and support philosophy, including a cost-benefit analysis of upgrade options.
Internal interfaces	Upgrade of CNRI to support the requirements of interface between the tactical trunk subsystem and combat radio subsystem;
	Provision of the interfaces between the subsystems of the tactical communications system, including the capacity of these interfaces;
	Limitations imposed by current technology on the basic requirements of interfaces;
	A detailed work breakdown structure and cost analysis for the provision of the interfaces;
	Manning, ILS, maintenance, and support philosophy, including a cost-benefit analysis of upgrade options.

Table 10.4 (continued)

Location	Issues
External interfaces	Provision of the interfaces between the tactical communications system and the strategic communications system, overlaid communications systems, and supported systems, including the capacity of these interfaces and the number required;
	Limitations imposed by current technology on the basic requirements of interfaces;
	The types of data that are to be carried by the interface;
	A detailed work breakdown structure and cost analysis for the provision of the subsystem;
	Manning, ILS, maintenance, and support philosophy, including a cost-benefit analysis of upgrade options.

Endnotes

[1] This concept of the local subsystem is similar to the LAS concept in BOWMAN. See, for example, "Bowman Watershed," *Military Technology*, Nov. 1998, pp. 51–53.

[2] ANSI/IEEE Std 802.11, "Wireless LAN Medium Access Control (MAC) and Physical Layer (PHY) Specifications," Piscataway, NJ: IEEE, 1999.

[3] "Specification of the Bluetooth System," Bluetooth SIG, Version 1.1, Feb. 2001.

[4] Policy for the GIG is defined in: U.S. Department of Defense Chief Information Officer Guidance and Policy, Aug. 24, 2000, Memorandum 10-8460, "GIG Network Operations," Memorandum 7-8170, "GIG Information Management," and Memorandum 4-8460, "GIG Networks."

[5] Mouly, M., and M. Pautet, *The GSM Systems for Mobile Communications*, Palaiseau: Cell and Sys, 1992.

List of Acronyms

2IC   second in command

ac   alternating current

ACN   airborne communications node

ADDS   advanced data distribution system

ADF   Australian Defence Force

ADFORM   Australian defense formatted message

ADPCM   adaptive differential pulse-code modulation

AF   audio frequency

AM   amplitude modulation

AO   area of operations

APC   armored personnel carrier

APCO   Association of Public-Safety Communications Officials International

ARQ   automatic repeat request

ASK   amplitude shift keying

ATACS   Army Tactical Communications System

ATM   asynchronous transfer mode

ATU   antenna tuning unit

AWACS   airborne warning and control system

BMA   brigade maintenance area

BS   base station

BSA   brigade support area

C2   command and control

C2W   command and control warfare

C3   command, control, and communications

C4ISR   command, control, and communications, computers, intelligence surveillance, and reconnaissance

CA   certificate authority

CCE   communications control element

CDMA   code-division multiple access

CELP   codebook-excited linear predictive

CIS   communications and information systems

CMA   corps maintenance area

CMC   communications management controller

CMM   communications control module

CNR   combat net radio

CNRI   combat net radio interface

CO   commanding officer

COMSEC   communications security

COTS   commercial off-the-shelf

CP   command post

CRC   cyclic redundancy check

CSA   corps support area

CSMA   carrier-sense multiple access

CSMA/CD   carrier-sense multiple access/collision detection

CVSDM   continuously variable slope delta modulation

CW   carrier wave

DARPA   Defense Advanced Research Projects Agency

dc   direct current

DCS   Defense Communications System

DCT   discrete cosine transform

DF   direction finding

DFT   discrete Fourier transform

DKLT   discrete Karhunen-Loeve transform

DMA   divisional maintenance area

DPSK   differential phase-shift keying

DSA   divisional support area

DSB   double sideband

DSBSC   double-sideband suppressed-carrier

DSSS   direct-sequence spread spectrum

DTSO   Defence Science and Technology Organisation (Australia)

DWHT   discrete Walsh-Hadamard transform

EA   electronic attack

EAC   echelons above corps

EDC   error detection and correction

EHF   extremely high frequency

ELF   extremely high frequency

EMC   electromagnetic compatibility

EP   electronic protection

EPLRS   Enhanced Position Locating and Reporting System

EPUU   EPLRS User Unit

ES   electronic support

ESM   electronic support measures

EW   electronic warfare

FAA Federal Aviation Administration

FCC Federal Communications Commission

FDM frequency-division multiplexing

FDMA frequency-division multiple access

FEC forward error correction

FELIN Fantassin à Equipment et Liason Integrées

FH frequency hopping

FIST Future Integrated Soldier Technology

FM frequency modulation

FSK frequency shift keying

GBS Global Broadcast System

GEO geostationary Earth orbit

GIG global information grid

GPS global positioning system

GSM Global System for Mobile Communication

HAAP high-altitude aeronautical platform

HALE high-altitude long-endurance

HALO High Altitude Long Operation

HAP high-altitude platform

HF high frequency

HQ   headquarters

IEEE   Institute of Electrical and Electronic Engineers

IF   intermediate frequency

IO   information operations

IP   Internet Protocol

IR   infrared

ISA   Italian Space Agency

ISB   independent sideband

ISDN   integrated subscriber digital network

ISO   International Standards Organization

ISP   Internet service provider

ITU   International Telecommunication Union

IW   information warfare

JTIDS   Joint Tactical Information Distribution System

JTRS   Joint Tactical Radio System

JV2020   Joint Vision 2020

KEA   key exchange algorithm

LAN   local area network

LAS   local area system

LEO   low-Earth orbit

LF   low frequency

LPC   linear predictive coding

LPD   low probability of detection

LPI   low probability of intercept

LSB   lower sideband

MAN   metropolitan area network

MBITR   Multiband Inter/Intra Team Radio

MEO   medium-Earth orbit

MF   medium frequencies

MOOW   military operations other than war

MSC   mobile switching center

MSE   Mobile Subscriber Equipment

MSS   mobile satellite service

MTF   message text format

NATO   North Atlantic Treaty Organization

NBC   nuclear, biological, and chemical

NCS   net control station

NCW   network-centric warfare

NEW   network-enabled warfare

NIC   network interface card

NMF   network management facility

NOC   node operations center

NOS   network operating system

NTDR   Near-Term Digital Radio

OODA   observation, orientation, decision, and action

OSI   open systems interconnection

OUDA   observation, understanding, decision, and action

PCM   pulse code modulation

PCS   personal communications systems

PDA   personal digital assistant

PLRS   Position Locating and Reporting System

PM   phase modulation

PN   pseudonoise

PPLI   precise participant location and identification

PSK   phase shift keying

PSTN   public-switched telephone network

QAM   quadrature amplitude modulation

QoS   quality of service

QPSK   quadrature phase-shift keying

RAP   radio access point

RATEL   radio telephone procedure

RF   radio frequency

RLC   run-length coding

RPE   regular pulse excited

RS   Reed-Solomon

SAR   synthetic aperture radar

SCRA   single-channel radio access

SDS   signal dispatch service

SF   special forces

SHF   super high frequency

SINCGARS   Single Channel Ground and Air Radio System

SOP   standing operating procedures

SSB   single sideband

STA   surveillance and target acquisition

STP   shielded twisted pair

TADIL   tactical data link

TBS   theater broadcast system

TCP   Transmission Control Protocol

TDM   time-division multiplexing

TDMA   time-division multiple access

TETRA   Terrestrial Trunked Radio

TIDP   Technical Interface Design Plan

TII   tactical interface installation

TOC   tactical operations center

UAV   uninhabited aerial vehicle

UHF   ultra high frequency

USB   upper sideband

USMTF   U.S. message text format

UTP   unshielded twisted pair

UWB   ultra wideband

VF   voice frequency

VHF   very high frequency

VLF   very low frequency

VoIP   voice over IP

VPN   virtual private network

WAN   wide area network

WIN   Warfighter Information Network

WLAN   wireless local area network

About the Authors

Dr. Michael J. Ryan received his B.E., M.Eng.Sc., and Ph.D. in electrical engineering from the University of New South Wales, Canberra, Australia, in 1981, 1989, and 1996, respectively. Since 1981, he has held a number of positions in communications and systems engineering and in management and project management as a lieutenant colonel with the Royal Australian Signal Corps. Since 1998, he has been with the School of Electrical Engineering at the University of New South Wales at the Australian Defence Force Academy, where he is currently a senior lecturer. His research and teaching interests are in communications systems (network architectures, electromagnetics, radio wave propagation, mobile communications, and satellite communications), information systems architectures, data compression for remote sensing applications, systems engineering, project management, and technology management. He is the editor-in-chief of the *Journal of Battlefield Technology* and has authored a number of articles on communications and information systems as well as a book on battlefield command systems. He has also coauthored books on communications and information systems, tactical communications electronic warfare, and systems engineering.

Dr. Michael R. Frater is an associate professor at the School of Electrical Engineering at the University of New South Wales at the Australian Defence Force Academy. He has more than 10 years of experience in the development of communications systems and services, including videoconferencing and video and image surveillance. He has led a number of collaborative projects investigating image and video communications over low-bandwidth links. Dr. Frater has been actively involved in the development of international standards for audio-visual communications and broadcasting and has served as head of the Australian delegation to the Moving Picture Expert Group, one of the major international standards bodies working in this area. He holds a B.Sc. and a B.E. in electrical engineering from the University of Sydney and a Ph.D. in systems engineering from the Australian National University. His research and teaching interests lie in digital audio-visual communications, including compression; transmission and delivery electronics; broadcasting and datacasting; telecommunications networks and architectures; and mobile communications (PCS technology and services). He has authored a number of articles on communications systems and communications services and coauthored books on digital television, communications and information systems, and tactical communications electronic warfare.

Index

The Artech House Information Warfare Library

For further information on these and other Artech House titles, including previously considered out-of-print books now available through our In-Print-Forever® (IPF®) program, contact:

Artech House	Artech House
685 Canton Street	46 Gillingham Street
Norwood, MA 02062	London SW1V 1AH UK
Phone: 781-769-9750	Phone: +44 (0)20-7596-8750
Fax: 781-769-6334	Fax: +44 (0)20-7630-0166
e-mail: artech@artechhouse.com	e-mail: artech-uk@artechhouse.com

Find us on the World Wide Web at:
www.artechhouse.com

Printed in the United States
124561LV00001B/118-156/A

9 781580 533232